Textual Data Science with R

Chapman & Hall/CRC
Computer Science and Data Analysis Series

The interface between the computer and statistical sciences is increasing, as each discipline seeks to harness the power and resources of the other. This series aims to foster the integration between the computer sciences and statistical, numerical, and probabilistic methods by publishing a broad range of reference works, textbooks, and handbooks.

Proposals for the series should be sent directly to one of the series editors above, or submitted to:

Chapman & Hall/CRC
Taylor and Francis Group
3 Park Square, Milton Park
Abingdon, OX14 4RN, UK

Semisupervised Learning for Computational Linguistics
Steven Abney

Visualization and Verbalization of Data
Jörg Blasius and Michael Greenacre

Chain Event Graphs
Rodrigo A. Collazo, Christiane Görgen, and Jim Q. Smith

Design and Modeling for Computer Experiments
Kai-Tai Fang, Runze Li, and Agus Sudjianto

Microarray Image Analysis: An Algorithmic Approach
Karl Fraser, Zidong Wang, and Xiaohui Liu

R Programming for Bioinformatics
Robert Gentleman

Exploratory Multivariate Analysis by Example Using R
François Husson, Sébastien Lê, and Jérôme Pagès

Bayesian Artificial Intelligence, Second Edition
Kevin B. Korb and Ann E. Nicholson

Computational Statistics Handbook with MATLAB®, Third Edition
Wendy L. Martinez and Angel R. Martinez

Exploratory Data Analysis with MATLAB®, Third Edition
Wendy L. Martinez, Angel R. Martinez, and Jeffrey L. Solka

Statistics in MATLAB®: A Primer
Wendy L. Martinez and MoonJung Cho

Clustering for Data Mining: A Data Recovery Approach, Second Edition
Boris Mirkin

Introduction to Machine Learning and Bioinformatics
Sushmita Mitra, Sujay Datta, Theodore Perkins, and George Michailidis

Introduction to Data Technologies
Paul Murrell

R Graphics
Paul Murrell

Correspondence Analysis and Data Coding with Java and R
Fionn Murtagh

Data Science Foundations: Geometry and Topology of Complex Hierarchic Systems and Big Data Analytics
Fionn Murtagh

Pattern Recognition Algorithms for Data Mining
Sankar K. Pal and Pabitra Mitra

Statistical Computing with R
Maria L. Rizzo

Statistical Learning and Data Science
Mireille Gettler Summa, Léon Bottou, Bernard Goldfarb, Fionn Murtagh, Catherine Pardoux, and Myriam Touati

Bayesian Regression Modeling With INLA
Xiaofeng Wang, Yu Ryan Yue, and Julian J. Faraway

Music Data Analysis: Foundations and Applications
Claus Weihs, Dietmar Jannach, Igor Vatolkin, and Günter Rudolph

Foundations of Statistical Algorithms: With References to R Packages
Claus Weihs, Olaf Mersmann, and Uwe Ligges

Textual Data Science with R
Mónica Bécue-Bertaut

Textual Data Science with R

Mónica Bécue-Bertaut

CRC Press is an imprint of the
Taylor & Francis Group, an **informa** business
A CHAPMAN & HALL BOOK

CRC Press
Taylor & Francis Group
6000 Broken Sound Parkway NW, Suite 300
Boca Raton, FL 33487-2742

First issued in paperback 2021

Version Date: 20190214

ISBN-13: 978-1-03-209365-9 (pbk)
ISBN-13: 978-1-138-62691-1 (hbk)

Publisher's Note
The publisher has gone to great lengths to ensure the quality of this reprint but points out that some imperfections in the original copies may be apparent.

Visit the Taylor & Francis Web site at
http://www.taylorandfrancis.com

and the CRC Press Web site at
http://www.crcpress.com

Contents

Foreword

This book by Mónica Bécue-Bertaut and her colleagues is original in a way that is not immediately noticeable because, as the reader will see, the work is modestly presented. The originality comes from bringing together developments from quite distinct domains, requiring a wide range of skills. It is uncommon indeed to find together in the same book rigorous theoretical results, methodological details of implementation, a real-world guide to the use of freely available software (here the R package **Xplortext**) and, lastly, several case studies based on collections of real textual data. Each of these case studies corresponds to a particular situation, dealt with right down to the computational details, which readers, progressively more autonomous, will be able to obtain themselves. These case studies cover a wide range of applications (medical databases, socio-economic surveys, sensory descriptions, political discourses), highlighting the versatility, richness, and limits of the tools, while simultaneously intriguing, motivating, and even entertaining the reader.

Alongside the presentation of classical results in standard correspondence analysis and clustering (results which reflect the author's experience in teaching and research), we also discover the more specialized hierarchical clustering with contiguity constraints, applied here in the context of chronological textual data, and the technique of multiple factor analysis for contingency tables, methods which Mónica Bécue-Bertaut has contributed to, following in the paths of Brigitte Escofier and Jérôme Pagès.

The structural analysis of textual data is an area in which Mónica Bécue-Bertaut has much contributed over the last twenty years, in terms of scientific publications, books, and software, as well as full-scale applications involving the processing of socio-economic survey data, political and legal text analysis, and the study of free text comments collected in the framework of sensory analyses. Other contributors to this book, especially in terms of applications, are experienced statisticians who have published extensively and have a high level of knowledge in the applied settings examined here.

An interdisciplinary field such as quantitative text analysis has a scientific status that is difficult to evaluate, due to its hybrid nature. The point of view taken by Mónica Bécue-Bertaut and colleagues in this book is one of a simplicity of process (without theoretical concessions) and pragmatism perfectly illustrating the popular adage: "The proof of the pudding is in the eating". Here, certainly, readers will acquire knowledge, but they will also easily obtain the ability to use specialized software on real data. They will be ready to be measured on the quality of the obtained results.

If the author's aim was to optimally pass on her wide-ranging and long-term knowledge of both these subjects and the technical tools that come with them, she has clearly succeeded.

Ludovic Lebart
Telecom-ParisTech

Preface

Textual data science focuses on the analysis of a set of documents, called a *corpus* in the linguistics setting, mainly using exploratory multivariate statistical methods. We are dealing with *textual data*, which is very different in nature to other quantitative and categorical data types used extensively in statistics and data science. Many methods have been imported from other fields; yet more have been designed expressly to deal with corpora. This is the case of *correspondence analysis* (CA), which is surprisingly little-known.

Though we often define the fields of application of statistics as those which can provide observed numerical values and potentially massive data sets, it is settled that such tools can also successfully extract information from texts. This is particularly necessary in fields that, by their nature, produce almost exclusively textual data that cannot be recoded into quantitative or qualitative variables.

In this list, first let us mention law and the justice system, which essentially deal in oral and written discourse, often stored in large legal databases. Second, bibliographical databases, which constitute a current challenge in technological watch. Third, one can go back to classical texts, which have frequently been examined and questioned by linguists and led to the design of specific methods. Fourth, the analysis of film scripts has brought new methodological points of view into the open. We also note the treatment of free text answers to open-ended questions in surveys as an important application of the field. This type of question facilitates the collection of information without any prior hypotheses. In fact, as can be perceived above, the application areas concerned by textual data science are all those that produce texts, from a small scale up to a very large one.

Computing tools are essential for the statistical analysis of large corpora. However, the overall approach remains fundamentally statistics-based. Documents are studied and compared by means of CA, starting from the whole set of words, their repetitions and distributions, without any prior hypotheses. As a result, the similarities between documents and between words, as well as associations between documents and words, can be visualized. Words that discriminate between documents can then be extracted, leading to the identification of important topics. In addition, the documents favoring these topics can be found and displayed. Grouping the documents into lexically homogeneous clusters by means of automatic processing is another focus of the subject and this book. Furthermore, simultaneous analysis of multilingual free text answers is made possible and straightforward, with no requirement of a prior translation.

Readers of the book

This book is intended for a wide audience, including statisticians, social science researchers, historians, literature analysts, linguistics experts, and essentially anyone facing the need to extract information from texts. It can also be a basic handbook for students currently training in data science, a field where the ability to process massive corpora is becoming essential.

Book content

Any corpus has to be converted into a table of documents × words. This initial encoding requires making essential choices. Chapter 1 addresses this issue in detail. Chapter 2 is devoted to the principles of CA in the framework of textual data. Chapter 3 presents applications of CA to a range of different example types. In Chapter 4, the clustering of documents previously encoded and mapped into a Euclidean space by CA is described and discussed. Chapter 5 then provides a methodology to identify words and/or documents characterizing parts of a corpus. This then allows us to summarize the textual content of the parts. Next, Chapter 6 sets out the principles of multiple factor analysis for contingency tables (MFACT), a more recent method dealing with multiple lexical tables, the applications of which are varied and wide-ranging. Examples include the analysis of multilingual corpora and assessing the impact of lemmatization. MFACT also tackles mixed data tables, bringing textual and contextual data together to enable the joint analysis of free text and closed answers in surveys. Lastly, Chapter 7 details the classical outline for reporting the results of any textual data analysis. Then, four real applications are developed in detail by statisticians, experts in textual data science. They deal with (i) a bibliographic database, work by Annie Morin, (ii) a unique rhetorical speech, by myself, (iii) a corpus of political speeches, by Ramón Álvarez-Esteban and myself, (iv) a peculiar short corpus collected in the framework of a sensory study, by Sébastien Lê and myself. Though these corpora come from quite different fields, each of them is associated with socio-economic repercussions of some kind. A short appendix, by Annie Morin, briefly presents the package **Xplortext** and gives some information about other textual data science packages.

The R package Xplortext

Examples in the book are processed with the help of the **Xplortext** package, an add-on to the open-source R software (https://cran.r-project.org). The databases and R scripts can be found at the website http://Xplortext.org. With these in hand, it will be easy to become familiar with the methods described in the book by reproducing all of the examples.

Acknowledgments

First and foremost, it is my pleasure to thank Ludovic Lebart for agreeing to write the preface of this work. I am also grateful to François Husson, editor of the French version, for his contributions and suggestions. I wish to acknowledge the help and assistance provided by David Grubbs, editor of this book at Taylor & Francis. Chapter 6 is mainly issued from working jointly with Jérôme Pagès. The rereadings by Bénédicte Garnier and Claude Langrand were greatly appreciated. In addition, I am very obliged to Daniele Castin and Kevin Bleakley for their contribution to the translation.

The package **Xplortext** was designed concurrently with the writing of both the French and English versions. Ramón Álvarez-Esteban has played a leading role in this work to which Belchin Kostov, Josep-Antón Sánchez and myself have collaborated.

I also want to thank my son, Javier, for his support and encouragement.

1

Encoding: from a corpus to statistical tables

1.1 Textual and contextual data

In the field of linguistics, a body of written or spoken documents is called a *corpus* (plural *corpora*). A large range of corpora, although differing in nature and reason for being, can be analyzed using the same exploratory multivariate statistical methods. This requires they be encoded as *two-way frequency tables* with a consistent structure. Furthermore, available contextual data can also be used to improve information extraction.

1.1.1 Textual data

The encoding step is particularly important in the field of textual data. Its aim is to convert a corpus into a *lexical table* of documents× words. Let us now clarify what we mean by *document* and *word*.

A corpus needs to be divided into *documents*, which correspond to the *statistical units* in the analysis. This division can sometimes be obvious but, at other times, choices have to be made. For example, when processing a corpus of free text answers collected by means of a questionnaire-based survey including open-ended questions, each individual answer is frequently considered as one document. Nevertheless, aggregation of the free text answers in terms of the values of a contextual qualitative variable into category documents is another possible choice (see Section 3.2). In the case of theater plays or film scripts, a division into scene documents is generally a good way to proceed. However, other options are possible. For instance, one of the first applications of correspondence analysis (CA), a core method in our conception of the statistical analysis of texts, was performed on Jean Racine's play *Phèdre*. There, Brigitte Escofier grouped together the dialogue of each character into separate documents. Another example is the analysis of the speech of the French Minister of Justice, Robert Badinter, championing the abolition of the death penalty in France in 1981. Here, the aim is to unveil the organizing principles of his argument, which are of particular importance in a rhetorical speech, i.e., one which aims to *convince* its audience. In this kind of analysis, it is necessary to divide the speech into rather similar-length sequences of text.

Each sequence, now looked at as a document, has to be long enough (from 500 to 1000 entries) to obtain a meaningful analysis when applying CA (see Section 7.3).

We will explain the difference between documents and what are known as *aggregate documents* in Section 1.1.3.

Word is used here as a generic term corresponding to the *textual unit* chosen, which could be for instance the *graphical form* (= continuous character strings), *lemmas* (= dictionary entries), or *stems* (= word roots). These are described further in Section 1.3. To count word frequencies first requires defining rules to segment a corpus into occurrences, and match each of them with a corresponding word.

When the documents have been defined, words identified, and counts performed, the lexical table to be analyzed can be constructed. This table contains the frequency with which the documents (rows) use the words (columns).

1.1.2 Contextual data

Contextual data is external information available on the documents, particularly important in the case of a corpus of free text answers collected through open-ended questions. Here, respondents have also answered many closed questions, later encoded in the form of quantitative or qualitative variables. However, any document can be linked to contextual data, which could be the publishing date, author, chronological position in the whole corpus, etc.

For example, in the case of plays or screenplays, each scene includes, besides dialogue, information such as character names, locations (outdoors, indoors, specific places, etc.), and a description of the action.

Contextual data are encoded as a standard documents × variables table called a *contextual table*.

1.1.3 Documents and aggregate documents

The same corpus can be divided into documents in different ways. The finest possible segmentation of the corpus into documents consists in considering each row of the original database as a document, called in the following a *source document*. In this case, an LT is constructed and analyzed. Then, if relevant, these documents can be aggregated according to either the categories of a contextual variable or clusters obtained from a clustering method, leading to construction of an *aggregate lexical table* (ALT). Aggregation of source documents into larger ones is particularly of interest when they are numerous and short, such as with free text answers, dialogue elements in a film script, and so on.

Nevertheless, aggregation can be carried out even if documents are long and/or not plentiful, depending on what the goals are. In any case, running analyses at different levels of detail may be useful. An analysis performed on source documents is referred to as a *direct analysis*, whereas *aggregate analysis* concerns aggregate documents. It must not be forgotten that the latter are aggregates of statistical units. As contextual variables are defined at the source document level, direct and aggregate analyses differ depending on the role the variables play in a given statistical method. Some indications will be provided in the following chapters.

1.2 Examples and notation

To illustrate the first six chapters of the book, we use part of the data collected during the *International Aspiration* survey, itself an extension of the *Aspiration* survey—which inquired about the lifestyles, opinions and aspirations of French people, to several countries. *Aspiration* was created by Ludovic Lebart at the *Centre de recherche pour l'étude et l'observation des conditions de vie* (CREDOC). Since the first wave in 1978, various open-ended questions have been introduced, leading Ludovic Lebart to develop original statistical methodology to address this type of textual data. Later, collaboration between Ludovic Lebart and Chikio Hayashi (then Director of the Institute of Statistical Mathematics in Tokyo) led them to design the *International Aspiration* survey, which included closed-ended questions from the original French survey, and open-ended questions such as:

- *What is most important to you in life?*
- *What are other very important things to you?* (relaunch of the first question)
- *What do you think of the culture of your country?*

International Aspiration was conducted in seven countries (Japan, France, Germany, the United Kingdom (UK), the United States (US), the Netherlands, and Italy) by Masamichi Sasaki and Tatsuzo Suzuki in the late 1980s. The questionnaire was translated into each of the respective languages. The respondents answered in their own language, thus providing the multilingual corpora *Life* (composed of the free text answers to the first two questions concatenated as if they were a single answer) and *Culture* (answers to the third question), both distinctive in the genre. These corpora, used for 25 years, have resulted in numerous publications and reference works. In Chapters 1–5, we use the UK component of both corpora (*Life_UK* and *Culture_UK*), and in Chapter 6, the UK, French and Italian components of the *Life* corpus (*Life_UK*, *Life_Fr*, and *Life_It*).

In this chapter, we use as an example the *Life_UK* corpus, collected from 1043 respondents. The contextual data consists of only one quantitative variable: `Age`, and seven qualitative variables: `Gender`, `Age_Group`, `Education`, pairwise `Gender_Age`, `Gender_Educ`, `Age_Educ`, and the triple `Gen_Edu_Age`.

TABLE 1.1
Excerpt from the raw corpus of *Life_UK*: free text answers of the first ten respondents. Elisions have been removed from the database.

Resp.	*"What is most important to you in life?"* *"What are other very important things to you?"*
1	good health; happiness
2	happiness in people around me, contented family, would make me happy; contented with life as a whole
3	contentment; family
4	health; happiness, money, family
5	to be happy; healthy, have enough to eat, enough money to live on
6	my wife; music, holidays, I like breaks, continuous good health
7	health; happiness
8	to be healthy; just to live long enough to see the children grow up; I do not think there is a lot to ask for, really
9	health; keeping going, family, going out, shopping, visiting
10	husband; new baby granddaughter, life in general

The `TextData` function of the R package **Xplortext** builds the LT from the raw corpus (see Table 1.1). This table has as many rows as there are non-empty documents, i.e., the number of respondents that answered ($I = 1043$), and as many columns as there are different words ($J = 1334$). The number of different words is called the *vocabulary size*. At the intersection between row i and column j, the (i, j)th entry contains the count y_{ij} of how many times document i uses word j. In the row and column margins of the LT we find, respectively, the column and row sums. The notation for a row/column margin term is obtained by substituting a dot for the index over which the summation is performed. Thus (see Figure 1.1):

$$y_{i\cdot} = \sum_{j=1}^{J} y_{ij} \qquad y_{\cdot j} = \sum_{i=1}^{I} y_{ij} \qquad N = y_{\cdot\cdot} = \sum_{i,j} y_{ij} \,.$$

The *column margin* is the column vector whose I terms correspond to the total number of occurrences (words) in each document. The *row margin* is the row vector whose J terms correspond to the number of times each word was used in the whole corpus. The grand total N corresponds to the total number of occurrences, also known as the *corpus size*.

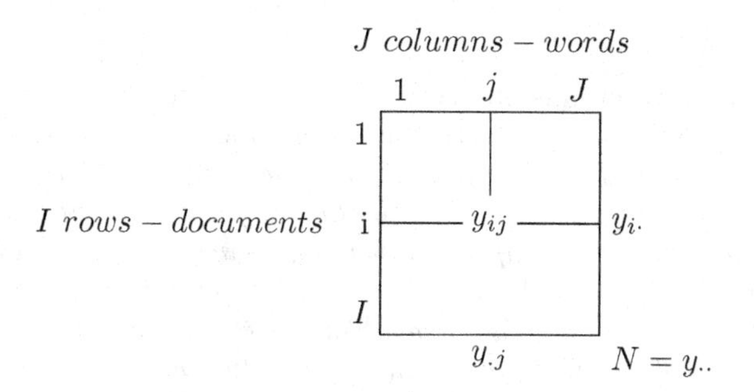

FIGURE 1.1
Lexical table **Y**; y_{ij} is the frequency with which document i uses word j; $y_{\cdot j}$ is the frequency of the word j in the corpus; $y_{i\cdot}$ is the length of document i; N is the corpus size.

This table can be turned into a relative frequency table, denoted **F**, by dividing each term by N. The (i, j)th entry of **F**, which has the same dimensions as **Y**, is given by $f_{ij} = y_{ij}/N$. The margins of **F** are denoted:

$$f_{i\cdot} = \sum_{j=1}^{J} f_{ij} \qquad f_{\cdot j} = \sum_{i=1}^{I} f_{ij} \qquad \sum_{i,j} f_{ij} = 1.$$

The contextual table, which we call **X**, is a documents × variables table with quantitative and/or qualitative variables. Each qualitative variable involves a certain number of categories.

The `TextData` function builds the LT and the contextual table. Words are included or discarded based on their frequency, the number of documents using them, and perhaps whether or not they are in a given *stoplist* (= list of words to be eliminated). Only non-empty documents are retained (non-empty after word selection), and certain documents may be removed by users, depending on their goals.

1.3 Choosing textual units

As noted before, *word* here is a generic term synonymous with *textual unit*, which encompasses different possible choices, detailed below.

1.3.1 Graphical forms

Aïno Niklas-Salminen has observed that the only rigorous definition of *word* results from its typographical transcription, and defines it as *"a sequence of letters, bounded to the right and left by spaces or punctuation marks"*. The *Larousse* dictionary considers it to be *"the language element consisting of one or several phonemes and capable of being graphically transcribed between two spaces"*. If we resort to the Oxford English Dictionary, *word* is defined as *"a single distinct meaningful element of speech or writing, used with others (or sometimes alone) to form a sentence and typically shown with a space on either side when written or printed"*.

Identifying words in terms of their spelling means choosing as textual unit the *graphical form*, or simply the *form* in the terminology adopted by André Salem. This corresponds to the three definitions given above and, further, has the great advantage of allowing simple automation of the identification and counting of occurrences of each word.

1.3.2 Lemmas

In a dictionary, words are presented in their *canonical forms*, called *lemmas* (in English: the infinitive in the case of verbs, masculine singular for nouns). In speech, lemmas appear in many forms, called *inflected forms* (mainly the conjugated form in the case of verbs, joining verbs and prepositions in the case of phrasal verbs, plural form in the case of nouns). Different inflected forms issued from the same lemma are generally written differently and thus correspond to different graphical forms. Inversely, different lemmas can give rise to inflected forms that are written similarly and thus match the same graphical form. For example, in the sentence, *The bandage was wound around the wound*, the first occurrence of the graphical form *wound* is a conjugated form of the verb *to wind*, while the second corresponds to the noun *wound*.

In order to perform an analysis on a *lemmatized* corpus (i.e., one in which each word has been substituted by its corresponding lemma), a morphosyntactic analyzer has to be used. This is a tool which transforms the succession of graphical forms that makes up a document into a succession of lemmas. A unique occurrence can correspond to two lemmas, as in the case of the abbreviation *don't,* with lemmas *to do* and *not.* Inversely, the two occurrences that form *going up* correspond to the unique lemma *to go up.* Identification of the lemma to which a graphical form corresponds is frequently context-dependent, as shown in the earlier example involving the word *wound.* There are numerous other examples of this. For instance, the graphical form *work* can correspond to the verb *to work*, or may be the canonical form of the noun *work.* It will be useful to differentiate between two lemmas with the same spelling but different grammatical categories, by attaching to each of them a label or *tag* indicating their category. Thus, for example, *minuteNoun* will be different from *minuteAdj.*

After processing, some ambiguity may remain. In any case, it should be borne in mind that any morphosyntactic analyzer produces a certain percentage of errors since, in the words of Ludovic Lebart, *"Issues with lemmatization show that no simultaneously reliable and fully automatable method can exist"*. Note also that these errors are not random, which can lead to distortions in the structure uncovered during a study.

Starting from either graphical forms or lemmas does not modify the steps in the analysis. The results obtained differ, of course, but in general lead to the discovery of similar structure. In Chapter 6, the same corpus is analyzed, successively, starting with each transcription: graphical forms and lemmas.

1.3.3 Stems

Stemmatization consists of grouping graphical forms with the same root into a single textual unit identified by this root or *stem*. Thus, in the *Life_UK* corpus, the graphical forms *health* and *healthy* could be grouped under the same stem *health*. Indeed, we can accept that *healthy* in the expression *to be healthy* conveys the same meaning as *health* does in *to have good health*.

1.3.4 Repeated segments

As is done by André Salem, we can also take into account *repeated segments*. Their definition is simple: any string of words not involving *strong punctuation marks* and repeated in the same way throughout a corpus is called a *repeated segment*. Note that in his terminology, strong punctuation marks are those which delimit repeated segments and longer sequences such as sentences, whereas *weak punctuation marks* delimit occurrences. In the R package **Xplortext**, default lists of strong and weak punctuation marks are provided; these can be modified by users.

In the *Life_UK* corpus, the most frequently repeated segments include *my family*, *good health*, *enough money*, and *peace of mind*, found respectively 236, 176, 49, and 42 times.

Identifying repeated segments and counting their frequency makes it possible to build a documents × repeated segments table.

1.3.5 In practice

The choice of textual unit depends on the application in mind, the goals of the study, and prior knowledge about the corpus, as well as on the availability of a morphosyntactic analyzer. The methods presented in this book apply in the same way regardless of the selected textual units. Nevertheless, we emphasize that repeated segments (compound textual units) are clearly distinct from the others. The segmenting of a corpus into graphical forms, lemmas, or stems, on the one hand, and repeated segments on the other, are different things. Indeed, a corpus can be seen as a sequence of word occurrences, while segmentation into repeated segments produces textual units which partially overlap. For instance, the four-word-long repeated segment *peace in the world* leads to the following two- and three-word-long segments: *peace in*, *in the*, *the world*, and *in the world*. The three-word-long segment *peace in the* does not appear in the list because it is a *constrained segment*, i.e., one which always appears in the same left and/or right context; here, *peace in* is always followed by the words *the world*. Chapters 3 and 7 will show how repeated segments contribute primarily as *supplementary elements* in CA.

Concerning the choice of textual unit, a well-known phrase can shed light on this discussion. Should we regroup under the same lemma (or the same stem) *reason* and *reasons* in *"the heart has its reasons which reason knows nothing of"*? This quote from the French author Pascal shows that the use of the singular or plural of the same lemma may convey important nuance.

This type of case can leave us in an inconclusive situation, essentially because there may be no universally best choice that can be made. Furthermore, it must be emphasized that the singular and plural forms of the same noun, or different forms of the same verb, will be close to or far from each other on CA plots, depending on their semantic similarity. So, to some extent, this makes lemmatization unnecessary and, in some cases, inappropriate. However, as CA requires eliminating infrequent words, certain forms will be dropped; whereas grouping occurrences based on the corresponding lemmas would have led us to take into account most of them. Nevertheless, language is sufficiently redundant that statistical analyses can usually tolerate these types of removal.

If lemmatization is possible, we recommend performing the same process on, successively, the non-lemmatized and lemmatized corpora. Generally, the results of both processes will not be very different, which can be seen as a kind of mutual validation. Any interpretation has to be mostly based on features common to both analyses. Correspondingly, if lemmatization is not feasible, the stability of results in the face of small perturbations, such as changing the frequency threshold, using or not using the words in the stoplist, etc., should be examined.

Note also that lemmatization provides not only lemmas corresponding to graphical forms, but also their grammatical or syntactic category. This type of information, including such things as the word index organized by syntactic category, frequency of each category, etc., can be very useful at other moments in an analysis.

In this work, the term *word* is used generically, regardless of the textual unit chosen by users. Various examples discussed in the following chapters will show the diversity of possible choices and their respective interest to us.

1.4 Preprocessing

1.4.1 Unique spelling

Counting the frequency of different words implies the identification of all occurrences. The correctness of these counts depends on the quality of the text transcription. Indeed, as is usual in statistics, the quality of results strongly depends on data quality.

In practice, users have to consider the potential value of making corrections to the text, prior to treatment. Such corrections mainly concern misspellings. It also needs to be ensured that each character has a unique status. For example, the period symbol used in acronyms and abbreviations like N.A.T.O. and U.N. is exactly the same as the one used to indicate the end of a sentence. For correct data treatment, these periods have to be removed and the words rewritten as NATO and UN.

Additional manual preprocessing steps can be considered. It may be useful to keep certain capital letters (those marking the beginning of proper names) but eliminate others (those beginning sentences). A compound word whose components are separated by a space can be identified as a single word provided that the space is suppressed; for example, *food security* can be rewritten as *foodsecurity*.

In the case of a corpus in English, it can sometimes be useful to eliminate elisions, i.e., convert *don't*, *isn't*, etc., into *do not*, *is not*, and so on, as we do in the example in Section 1.6.

Such preliminary corrections and/or modifications may require manual intervention on the corpus using a text editor and possibly its spell checker. If the corpus size makes this preprocessing too time consuming, we can instead run a quality measure on a sample extracted from the corpus.

1.4.2 Partially automated preprocessing

Some preprocessing steps have been automated in the **Xplortext** package. In particular, it is possible to:

- Request that the whole corpus be converted into lowercase (respectively, uppercase) letters, keeping the existing accents and other diacritics essential to the given language.

- Define symbols to be considered as punctuation marks—all others are then considered letters.

- Automatically eliminate numbers.

These operations are carried out by the `TextData` function, which was used to create the LT we will analyze in Section 1.7.

1.4.3 Word selection

The LT that is built for an analysis is called *full* if all unique words are kept, or *reduced* if only a part of them are. We can choose to keep only *semantic words*, i.e., words whose meaning is autonomous (mainly nouns, verbs, and adjectives), which are also known as *full words*. Keeping only *function words*, also known as *grammatical words*, *empty words* or *tool words* (mainly articles, conjunctions and prepositions), is another option. In fact, there is no clear distinction between the two types. For example, adverbs are attached to one category or the other, depending on the analysis. Word selection depends on factors such as prior knowledge about the corpus, methods to be applied, and the study goals.

Minimum thresholds are generally imposed on both word frequency and the number of documents in which a word has to be mentioned, in order to be retained. This allows for statistically significant comparisons to be made between documents.

Some software packages offer predefined stoplists of so-called *stopwords*. These lists, which are language-dependent, allow for systematic elimination of stopwords in subsequent analyses. **Xplortext** uses, on request, the stoplist available in the **tm** package. A personalized stoplist can also be used.

1.5 Word and segment indexes

The first output, after identifying various words and counting their frequencies, is the *word index*, in either lexicographic order (dictionary order) or lexicometric order (decreasing frequency), possibly limited to the most frequent words. Likewise, a *repeated segment index*, according to both possible orders, is output. The lexicographic ordering gives a visual idea of how two-word-long segments extend to three-word-long ones, four-word-long ones, etc.

1.6 The *Life_UK* corpus: preliminary results

1.6.1 Verbal content through word and repeated segment indexes

An examination of the full word and segment indexes, more extensive than those shown here in Tables 1.2 and 1.3, provides initial information about the verbal content of the documents. In this example, content is mainly conveyed by nouns. There is only a small number of adjectives. As for the verbs, they only show their full meaning in the context of other words.

TABLE 1.2
Index of the twenty most frequent words.

```
TextData summary

                          Before     After
Documents                1043.00   1040.00
Occurrences             13917.00  13917.00
Words                    1334.00   1334.00
Mean-length                13.34     13.38
NonEmpty.Docs            1040.00   1040.00
NonEmpty.Mean-length       13.38     13.38

Index of the  20  most frequent words
        Word Frequency N.Documents
1  my              810         461
2  family          705         624
3  health          612         555
4  to              523         297
5  and             504         375
6  the             332         233
7  of              312         255
8  good            303         254
9  a               300         222
10 i               287         184
11 happiness       229         218
12 in              181         157
13 money           172         169
14 life            161         149
15 that            160         128
16 job             143         137
17 is              141         118
18 happy           137         123
19 be              136         115
20 children        131         129
```

Family and *health* are the main subjects. The first is alluded to through *family*/705 (occurrences), *children*/131, *husband*/96, *wife*/76, *family life*/44, *grandchildren*/30, *daughter(s)*/21+5, *kids*/16, *marriage*/14, *parents*/13 and *welfare of my family*/4. As for health, we find *health*/612, *good health*/176, *healthy*/46, *my health*/48, while *health of my family*/5 corresponds to both subjects. Somewhat less cited but nevertheless important issues include *happiness* (*happiness*/229, *happy*/137, *contentment*/31, *happy family life*/13), *money and welfare* (*money*/172, *standard of living*/30, *welfare*/22, *enough money to live*/16), and *work* (*job*/143, *work*/118, *employment*/20, *job satisfaction*/13 and *career*/11). We also find words and segments related to general subjects like *peace* (included in both the *peace of mind* and *peace in the world* segments), *security*, and *freedom.* At the bottom of the list, if we only consider words used at least 10 times, secondary topics appear such as *holidays, leisure, religion,* and *social life.* Sometimes, we expect certain topics/words to be common but then observe that they were relatively infrequent or even absent, which in itself is also valuable information. For instance, moral qualities (*decent*/10) and education (*education*/25) rarely come to mind for UK inhabitants when the question of what is important in life is asked.

Repeated segments give partial information about the context of words, and thus provide information on their meaning. For example, we can better understand the role of certain occurrences of *to want* and *to do* when reading the segment *to do what i want.*

TABLE 1.3
Twenty segments selected from the 44 with at least 4 words and 4 repetitions.

Number of repeated segments 44

Index of the repeated segments in alphabetical order

	Segment	Frequency	Long
1	a good standard of living	6	5
2	able to live comfortably	5	4
4	be able to live	5	4
5	being able to live	4	4
6	can not think of anything	9	5
7	can not think of anything else	7	6
8	enough money to keep	5	4
9	enough money to live	16	4
10	enough money to live comfortably	4	5
14	good health for all	4	4
15	good standard of living	10	4
16	happiness of my family	4	4
18	having enough money to live	4	5
19	health of my family	5	4
22	i do not know	6	4
23	i do not think	4	4
29	my family s health	6	4
36	peace in the world	6	4
43	to do what i want	4	5
44	welfare of my family	4	4

In the case of free text answers, these initial results can strongly influence the following steps in the analysis.

Remarks:

The word "*I*" is written as "*i*" because the whole corpus is converted to lowercase. "*family's*" is written as "*family s*" because the apostrophe is removed just like the punctuation marks.

1.6.2 Univariate description of contextual variables

A univariate description of contextual variables can complement the word and segment indexes. In the case of a qualitative variable, the number of documents by category can be given. In the case of a quantitative one, a standard summary can be made (see Table 1.4).

TABLE 1.4
Univariate description of contextual variables.

```
Summary of the contextual categorical variables
   Gender       Age_Group        Education     Gender_Age
 Man  :495    >70     :136    E_Low   :479    M<=30 :134
 Woman:545    30_34   :104    E_Medium:418    M31_55:194
              20_24   :102    E_High  :143    M>55  :167
              25_29   : 98                    W<=30 :132
              35_39   : 93                    W31_55:251
              40_44   : 92                    W>55  :162
              (Other):415
    Gender_Educ             Age_Educ            Gen_Edu_Age
 M_EduLow :222    >55_Low        :235   W_Low_31_55:134
 M_EduMed :197    31_55_Low      :226   M_Low_>55  :121
 M_EduHigh: 76    <=30_Medium    :187   W_Low_>55  :114
 W_EduLow :257    31_55_Medium:159      M_Med_<=30 : 94
 W_EduMed :221    >55_Medium     : 72   W_Med_<=30 : 93
 W_EduHigh: 67    <=30_High      : 61   M_Low_31_55: 92
                  (Other)        :100   (Other)     :392

Summary of the contextual quantitative variables
      Age
 Min.    :18.00
 1st Qu.:30.00
 Median :44.00
 Mean    :45.81
 3rd Qu.:61.00
 Max.    :90.00
```

1.6.3 A note on the frequency range

The distribution of the counts V_f of the words with frequency f, $f = 1, \ldots, F_{max}$, in a given corpus is called the *frequency range* range.

Frequency:	1	2	3	4	5	...	F_{max}
Count:	V_1	V_2	V_3	V_4	V_5	...	$V_{F_{max}}$

Given that V and N correspond respectively to the vocabulary size and corpus size, we have that $\sum_{f=1}^{F_{max}} V_f = V$ and $\sum_{f=1}^{F_{max}} V_f \cdot f = N$. Although each corpus is associated with a specific distribution, frequency ranges tend to exhibit similar features across corpora. The most well-known example is Zipf's law, which says that the product between word frequency and word rank, when words are ordered by decreasing frequency, is approximately constant. Given that there are, at the same time, many infrequent words and very few highly occurring words, a Pareto chart is often used to represent the frequency range. Both of its axes have a logarithmic scale. The vertical axis corresponds to the frequency f, and the horizontal one to the number of words repeated at least f times in the corpus, denoted N_f (with $N_f = \sum_{i=f}^{Fmax} V_i$). In Figure 1.2, in accordance with Zipf's law, we see that the points from the *Life_UK* study are indeed approximately found on a straight line, except for the low frequency ones. We do not take a deeper look here at laws governing the frequency range. This is the subject of the field known as *stylistics*.

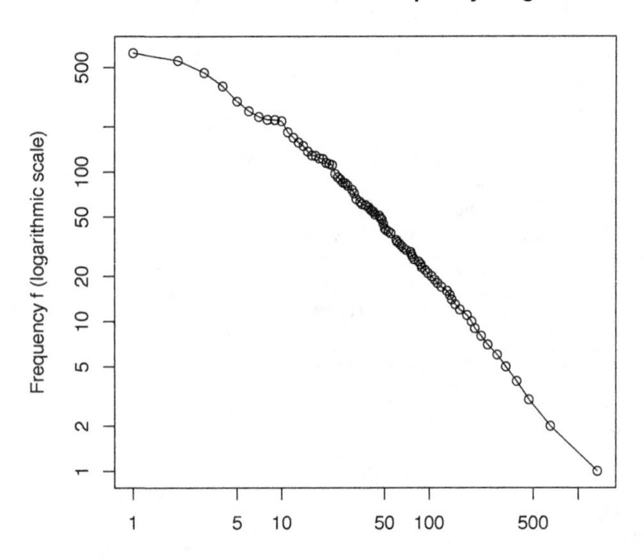

FIGURE 1.2
Life_UK corpus: the frequency range.

1.7 Implementation with Xplortext

This section shows how to obtain the results discussed in this chapter. The database is first imported using the script called `script.Importer.Aspi_UK.R`, starting from the `Aspiration_Int_UK_En.csv` database, which contains both the free text answers and the contextual variables. The first row contains the variable names, and the first column contains the document identifier (= respondent identifier). R commands are used to order the categories of ordinal variables.

Then, the `Aspiration_Int_UK_En.RData` database is created. Following this, the script `script.chap1.R` is used to obtain the results described as we moved through the chapter, involving the following steps:

- The *Life_UK* corpus corresponds to columns 9 and 10 in the `Aspiration_Int_UK_En.RData` database.
- The `TextData` function selects documents and words. In this example, all documents are kept, as well as all words. The LT is created in the form of the `$DocTerm` dataframe (compressed format).
- The `summary.TextData` function informs us about the size of the corpus and the vocabulary. It also provides word and segment indexes (see Tables 1.2 and 1.3). In both cases, the indexes are limited to the most frequent elements. Standard summaries of the selected contextual variables are also provided (see Table 1.4).
- Various R commands help us represent the frequency range using a Pareto diagram (see Figure 1.2).

1.8 Summary

Here are the main points to remember:

- The statistical methods we use analyze not the raw corpus but the lexical table, either LT or ALT, in which it is encoded. Therefore, building this table is an essential step in reflecting the viewpoint adopted by the user, who needs to have an understanding of what their encoding choices imply.
- Different encodings can be performed and the corresponding results compared.
- The best way to divide a corpus into documents is not always immediately obvious. The user's point of view must guide this, taking into account that documents will be considered as the statistical units in the analyses.
- The choice of the textual unit (graphical form, lemma, or stem) rarely has a huge influence on the structure of the results.
- The lexical table, either LT or ALT, resulting from the chosen encoding is a documents $\times$ words table.

- Depending on the goals, one (or more) ALTs can be built by aggregating documents with respect to values of one (or more) contextual qualitative variables.
- Contextual data plays an important role in the analysis and interpretation of results. This should be carefully recorded and encoded.

2

Correspondence analysis of textual data

2.1 Data and goals

2.1.1 Correspondence analysis: a tool for linguistic data analysis

Correspondence analysis (CA), as a method to analyze linguistic data, was proposed by Jean-Paul Benzécri who stated that it was able *"to address any type of issue related to the form, meaning or style of texts"*. The method rests on a eigendecomposition, which connects it to previous works on classical rectangular tables of positive numbers. Ludovic Lebart has extended the scope of this research to free text answers, leading him to conceive of the joint analysis of textual and contextual data.

2.1.2 Data: a small example

In this chapter, we use a small lexical table constructed from the *Life_UK* corpus to set out the principles of CA. Six documents are formed by aggregating the free text answers according to the six categories of the `Gender_Educ` variable (*men/women, low education*: 222 and 257 respondents, *men/women, medium education*: 197 and 221 respondents, and *men/women, high education*: 76 and 67 respondents). In this first example, a very high threshold on the word frequency is chosen with a view to helping us understand CA, step by step, applied to a low-dimensional table. Thus, only the nouns used at least 90 times are selected, excluding *husband* (96 times) (see Table 2.1). If we do not remove *husband*, we are forced to keep *wife* too (only 76 times), and both words would then be too influential on the results since each of them is used by only one gender. We finally obtain a lexical table with six row documents and ten column words (see Table 2.2).

The selected part of the corpus consists of $N = 2477$ occurrences of 10 distinct words. The lexical table, denoted $\mathbf{Y}$, has $I = 6$ document categories (in rows) and $J = 10$ words (in columns). The (i, j)th entry y_{ij} corresponds to the *frequency* with which document i uses word j. For example, the *women with medium education* document (composed of all free text answers given by this category) uses the word *family* 160 times. Terms $y_{i.}$ in the *column margin* indicate the length of each document. The *women with low education*

TABLE 2.1
Excerpt of the `TextData` summary: corpus size, vocabulary size and index of the most frequent words.

```
> summary(res.TD,nword=10)

                        Before   After
Documents              1043.00    6.00
Occurrences           13917.00 2477.00
Words                  1334.00   10.00
Mean-length              13.34  412.83
NonEmpty.Docs          1040.00    6.00
NonEmpty.Mean-length     13.38  412.83

Glossary of the words
        Word Frequency N.Documents
1  family          705           6
2  health          612           6
3  happiness       229           6
4  money           172           6
5  life            161           6
6  job             143           6
7  children        131           6
8  work            118           6
9  friends         116           6
10 home             90           6
```

document is the longest with 636 occurrences. Terms $y_{.j}$ in the *row margin* correspond to the *frequency* of each word in the whole corpus being analyzed. The word *family* is the most used with 705 occurrences.

TABLE 2.2
Table of documents × words and its margins.

	children	family	friends	happiness	health	home	job	life	money	work	Sum
M_EduLow	21	148	9	38	138	13	23	48	33	34	505
M_EduMed	14	121	26	47	99	19	42	34	38	27	467
M_EduHigh	4	42	12	16	32	3	20	9	9	13	160
W_EduLow	45	191	29	59	179	28	13	36	38	18	636
W_EduMed	40	160	24	57	138	19	31	27	44	19	559
W_EduHigh	7	43	16	12	26	8	14	7	10	7	150
Sum	131	705	116	229	612	90	143	161	172	118	2477

2.1.3 Objectives

The main goal in applying CA to a documents × words table is to visualize the proximity between documents, the proximity between words, and associations between documents and words. Two documents are close if, regardless of their length, they favor or avoid the same words. Two words are close if they are distributed in the same way across all of the different documents, regardless of their frequency. Lastly, we say that a category and a word *attract* (resp. *repel*) each other if the category uses the word with a higher (resp. lower) relative frequency than the average. Below, we specify the meaning of *proximity*, *attraction*, and *repulsion* in the CA context.

2.2 Associations between documents and words

2.2.1 Profile comparisons

Here, *row* and *document* are used as synonyms, as are *column* and *word*. Furthermore, *document category* (or *category*) refers to the document composed of all free text answers belonging to the corresponding category in the case of a corpus of free text answers.

If the strength of the association between document categories and words is evaluated from the raw frequencies, the *men with medium education* document category appears to be the one the most concerned by *job*, using this word the most (42 times). However, this word is relatively more important for *men with high education* who dedicate 20/160 = 12.50% of their occurrences to this word, as opposed to *men with medium education*, with 42/467 = 8.99%.

CA adopts this viewpoint and compares documents using the conditional distribution of words in each of them, which is either called the *row profile* or *lexical profile*.

These lexical profiles are computed from the lexical table either in frequency form **Y** (raw count table) or in relative frequency form **F** by dividing each row by its total. Row profile i is given by: $y_{ij}/y_{i\cdot} = f_{ij}/f_{i\cdot}$, $\{j = 1, \ldots, J\}$. Computing the average row profile $y_{\cdot j}/N = f_{\cdot j}$, $\{j = 1, \ldots, J\}$, allows us to compare the lexical profiles of the documents to the average, thus revealing categories using certain words more/less than the average. For example, *men with high education* use *job* (12.50% of occurrences) more than the average (5.77% of all occurrences), and is said to *attract* it. Inversely, *women with low education* use the word *job* less than average (2.04% of occurrences), and are said to *repel* it.

Similarly, words can be compared using their *column profiles*, computed from the frequency table **Y** (or relative frequency table **F**), by dividing each column by its total. The column profile j is given by: $y_{ij}/y_{\cdot j} = f_{ij}/f_{\cdot j}, \{i = 1, \ldots, I\}$. Computing the average column profile $y_{i\cdot}/N = f_{i\cdot}, \{i = 1, \ldots, I\}$, then allows us to compare a word's profile to the average, thus identifying words used more/less than average in a document. Both the row and column profile tables are shown in Table 2.3, along with the average row and column profiles (also called *marginal profiles*).

TABLE 2.3
Document row profiles and word column profiles.

Row profiles (in percentage)

	children	family	friends	happiness	health	home	job	life	money	work	Sum
M_EduLow	4.16	29.31	1.78	7.52	27.33	2.57	4.55	9.50	6.53	6.73	100
M_EduMed	3.00	25.91	5.57	10.06	21.20	4.07	8.99	7.28	8.14	5.78	100
M_EduHigh	2.50	26.25	7.50	10.00	20.00	1.88	12.50	5.62	5.62	8.12	100
W_EduLow	7.08	30.03	4.56	9.28	28.14	4.40	2.04	5.66	5.97	2.83	100
W_EduMed	7.16	28.62	4.29	10.20	24.69	3.40	5.55	4.83	7.87	3.40	100
W_EduHigh	4.67	28.67	10.67	8.00	17.33	5.33	9.33	4.67	6.67	4.67	100
Row av.-profile	5.29	28.46	4.68	9.25	24.71	3.63	5.77	6.50	6.94	4.76	100

Column profiles (in percentage)

	children	family	friends	happiness	health	home	job	life	money	work	C.av.-pr.
M_EduLow	16.03	20.99	7.76	16.59	22.55	14.44	16.08	29.81	19.19	28.81	20.39
M_EduMed	10.69	17.16	22.41	20.52	16.18	21.11	29.37	21.12	22.09	22.88	18.85
M_EduHigh	3.05	5.96	10.34	6.99	5.23	3.33	13.99	5.59	5.23	11.02	6.46
W_EduLow	34.35	27.09	25.00	25.76	29.25	31.11	9.09	22.36	22.09	15.25	25.68
W_EduMed	30.53	22.70	20.69	24.89	22.55	21.11	21.68	16.77	25.58	16.10	22.57
W_EduHigh	5.34	6.10	13.79	5.24	4.25	8.89	9.79	4.35	5.81	5.93	6.06
Sum	100.00	100.00	100.00	100.00	100.00	100.00	100.00	100.00	100.00	100.00	100.00

CA acts as a visual synthesis of the distributional similarities/dissimilarities of rows and columns by comparing all of the document/word profiles with one another, but also with the average row/column profile. In the reference situation, the row/column profiles are all equal to each other and also equal to the corresponding average profile. In other words, all of the documents have the same verbal content and all of the words have the same distribution across all of the documents.

2.2.2 Independence of documents and words

Another approach to study the associations between documents and words, equivalent to the one above, can be developed in terms of *departure from independence*. Here, the reference situation is the absence of relationships between documents and words, in the sense that the former select the latter at random. Let us begin with the lexical table in relative frequency form, denoted **F**. There is independence between the row documents and column words if, for any row i and column j, the following equation holds:

$$f_{ij} = f_{i\cdot} \cdot f_{\cdot j}.$$

In this case, the following equations can also be tested:

- For every row $i, \{i = 1, \ldots, I\}$, $f_{ij}/f_{i\cdot} = f_{\cdot j}, \{j = 1, \ldots, J\}$.
- For every column $j, \{j = 1, \ldots, J\}$, $f_{ij}/f_{\cdot j} = f_{i\cdot}, \{i = 1, \ldots, I\}$.

These formulas holding would mean that all lexical profiles were equal to one another and equal to the average row profile . Similarly, all column profiles would be equal to one another and equal to the average column profile. This clearly shows that both approaches, either the one developed here (departure from independence between documents and words) or the earlier one (comparison of document/word profiles), are equivalent. Nevertheless, each point of view helps us to focus on different and complementary aspects.

Of course, remember that total independence between documents and words is unlikely to be ever observed in a real corpus. However, it acts as a useful reference case here.

Under the *independence hypothesis*, i.e., assuming that $f_{ij} = f_{i\cdot} \cdot f_{\cdot j}$ holds for any i and j, the *expected relative frequencies* are given by $f_{i\cdot} \times f_{\cdot j}$, and the *expected counts* are $N \times f_{i\cdot} \times f_{\cdot j}$. The table of these expected counts is taken as the *model* to compare the table of observed counts $\mathbf{Y}$ with. This model is called the *independence model*.

TABLE 2.4
Tables of observed and expected counts.

Observed counts

	children	family	friends	happiness	health	home	job	life	money	work	Sum
M_EduLow	21	148	9	38	138	13	23	48	33	34	505
M_EduMed	14	121	26	47	99	19	42	34	38	27	467
M_EduHigh	4	42	12	16	32	3	20	9	9	13	160
W_EduLow	45	191	29	59	179	28	13	36	38	18	636
W_EduMed	40	160	24	57	138	19	31	27	44	19	559
W_EduHigh	7	43	16	12	26	8	14	7	10	7	150
Sum	131	705	116	229	612	90	143	161	172	118	2477

Expected counts

	children	family	friends	happiness	health	home	job	life	money	work	Sum
M_EduLow	26.7	143.7	23.6	46.7	124.8	18.3	29.2	32.8	35.1	24.1	505
M_EduMed	24.7	132.9	21.9	43.2	115.4	17.0	27.0	30.4	32.4	22.2	467
M_EduHigh	8.5	45.5	7.5	14.8	39.5	5.8	9.2	10.4	11.1	7.6	160
W_EduLow	33.6	181.0	29.8	58.8	157.1	23.1	36.7	41.3	44.2	30.3	636
W_EduMed	29.6	159.1	26.2	51.7	138.1	20.3	32.3	36.3	38.8	26.6	559
W_EduHigh	7.9	42.7	7.0	13.9	37.1	5.5	8.7	9.7	10.4	7.1	150
Sum	131.0	705.0	116.0	229.0	612.0	90.0	143.0	161.0	172.0	118.0	2477

The meaning and usage of this independence model is far from what is usual in statistics. In applied statistics, a model is expressed as a mathematical formula accounting for the relationships between variables in a simplified but interpretable way. The formula is given *a priori*, based on hypotheses on the relationships, and then compared—using tests—to collected data. This enables an evaluation of the model's relevance, i.e., its ability to predict or explain. Here, it is completely different: the question is not whether the independence model is relevant or not. Rather, what is important here is to see if we are sufficiently far enough away from independence that interesting textual features may potentially be gleaned from the data in later steps.

As an example of information that can be gained by a comparison of observed and expected count tables, let us return to the relationships between *women with low education* and *job*, and *men with high education* and *job*. As seen earlier, when comparing the profiles of these two document categories to the average profile, we concluded that the first category repels the word *job*, while the second attracts it. In fact, a comparison of the observed and expected counts leads to the same conclusion as the comparison of the row (resp. column) profile with its associated average profile. The first (resp. second) category uses the word *job* 13 (resp. 20) times, much less (resp. more) than under the independence model's expected count of 36.7 (resp. 9.2)—see Table 2.4.

CA studies all of the relationships between documents and words in terms of deviation from the independence model, and provides an interpretable geometric representation of these relationships. This then leads to the construction of what are known as *factorial axes*, which are described in Sections 2.3 and 2.4.

2.2.3 The χ^2 test

In the case of a real corpus, the independence model is never totally satisfied, and we need to evaluate the level of departure from it that has occurred. The χ^2 *statistic* summarizes the differences between observed and expected counts as follows:

$$\chi^2 = \sum_{i,j} \frac{(Nf_{ij} - Nf_{i\cdot} \times f_{\cdot j})^2}{Nf_{i\cdot} \times f_{\cdot j}} = N \sum_{i,j} \frac{(f_{ij} - f_{i\cdot} \times f_{\cdot j})^2}{f_{i\cdot} \times f_{\cdot j}}.$$

It is clear that the value of this statistic, i.e., this summary of the data, will be:

- Zero if the observed and expected counts are all equal.
- Small if they are close to being equal.
- Larger and larger, the further they are from being equal.

The first two possibilities are not very interesting situations, since all documents would have (approximately) the same *verbal content.*

A classical statistical test known as the χ^2 *test* uses this statistic to evaluate whether the level of departure from independence model is large enough to suggest non-independence, and thus whether or not it may be worth running

TABLE 2.5
Association rates between documents and words.

Association rates

	children	family	friends	happiness	health	home	job	life	money	work
	Terms									
Docs	children	family	friends	happiness	health	home	job	life	money	work
M_EduLow	0.79	1.03	0.38	0.81	1.11	0.71	0.79	1.46	0.94	1.41
M_EduMed	0.57	0.91	1.19	1.09	0.86	1.12	1.56	1.12	1.17	1.21
M_EduHigh	0.47	0.92	1.60	1.08	0.81	0.52	2.17	0.87	0.81	1.71
W_EduLow	1.34	1.06	0.97	1.00	1.14	1.21	0.35	0.87	0.86	0.59
W_EduMed	1.35	1.01	0.92	1.10	1.00	0.94	0.96	0.74	1.13	0.71
W_EduHigh	0.88	1.01	2.28	0.87	0.70	1.47	1.62	0.72	0.96	0.98

CA. Applying this test to lexical tables (either LT or ALT) (as if they were contingency tables) is not entirely legal from a theoretical point of view, due to the non-independence between occurrences found in the same document. However, more important than the result of this test, which in the case of a real corpus practically always leads to rejecting the independence hypothesis, our actual interest is in the relationship between the χ^2 statistic and the inertia of both the document and word clouds, denoted by Φ^2 and examined in Section 2.3.3.

In our running example, we obtain $\chi^2 = 133.5$. The associated p-value (here, $p = 1.1 \times 10^{-10}$) corresponds to the *statistical significance* of the result; the smaller the p-value, the larger the significance. Here, the p-value is very small, much lower than 0.05, the usual threshold in practice. This leads us to reject the independence hypothesis between documents and words. Consequently, the documents have different verbal content, words are associated in different ways in them, and certain documents favor specific words (attraction) and avoid others (repulsion).

2.2.4 Association rates between documents and words

The ratio of the observed count with respect to the expected count for a given row point (document i) and column point (word j) measures the association between this document and this word. This ratio τ_{ij} is called the *association rate* between document i and word j, and is given by:

$$\tau_{ij} = \frac{N \times f_{ij}}{N \times f_{i\cdot} \times f_{\cdot j}} = \frac{f_{ij}}{f_{i\cdot} \times f_{\cdot j}}.$$

The association rate is greater than 1 if the document and word attract each other, and less than 1 if they repel each other.

2.3 Active row and column clouds

The table we are interested in is called the *principal table*. The rows and columns of this table are said to be *active* because they are involved in computing the *factorial axes* (see Section 2.4.1 below). These rows, on the one hand, and columns, on the other, need to form homogeneous sets so that induced distances are meaningful. The results obtained from the principal table can be enriched using supplementary rows and/or columns (see Section 2.6).

2.3.1 Row and column profile spaces

The sets of both row documents and column words are taken into account via their profiles (see Table 2.3). CA looks for a geometric representation of the similarities and differences between profiles.

The document cloud N_I sits in a J-dimensional space called the *document space* whose orthogonal axes correspond to words. Each document i is represented by a point whose coordinates are the elements of its profile $f_{ij}/f_{i\cdot}$, $\{j = 1, \ldots, J\}$. In addition, document i is attributed a weight $f_{i\cdot}$, corresponding to the proportion its occurrences represent with respect to the grand total: $f_{i\cdot} = \sum_{j=1}^{J} y_{ij}/N$. The center of gravity (CoG) of this cloud, denoted G_I, has for profile the average document profile $f_{\cdot j}$, $\{j = 1, \ldots, J\}$, equal to the row margin of the table $\mathbf{F}$.

Similarly, the word cloud N_J sits in an I-dimensional space called the *word space* whose orthogonal axes correspond to documents. Each word is represented by a point whose coordinates are the elements of its profile $f_{ij}/f_{\cdot j}$, $\{i = 1, \ldots, I\}$. In addition, word j is attributed a weight $f_{\cdot j}$ corresponding to the proportion its occurrences represent in the grand total: $f_{\cdot j} = \sum_{i=1}^{I} y_{ij}/N$. The CoG of this cloud, denoted G_J, has for profile the average word profile $\{f_{i\cdot}, i = 1, \ldots, I\}$, equal to the column margin of the table $\mathbf{F}$.

We therefore have two point clouds of profiles sitting in two different spaces, with weights attributed to the points. The respective CoGs correspond to the margins of the relative frequency table $\mathbf{F}$. Both spaces now need to be endowed with distances in order to induce proximities between either rows or columns.

2.3.2 Distributional equivalence and the χ^2 distance

The document space (respectively, word space) needs to be endowed with a distance defined on the profiles so that the distance between two rows i and i' (resp. two columns j and j') is zero when the associated profiles are equal, and small when they are similar, whatever their relative lengths $f_{i\cdot}$ and $f_{i'\cdot}$ (resp. relative frequencies $f_{\cdot j}$ and $f_{\cdot j'}$).

Distances should also satisfy the *distributional equivalence* principle, which says that the distance between documents i and i' does not change when words j and j', with identical profiles, are merged. In the same way, we require that the distance between words j and j' does not change when documents i and i', with identical profiles, are merged. In such cases involving what are called *distributional synonyms*—here documents (resp. words), we merge the rows (resp. columns) they correspond to in the table $\mathbf{Y}$ as a single row (resp. column), equal to their sum, and thus proportional to them. The distributional equivalence principle and the requirement for a quadratic formula lead to the choice of the following distances between rows and columns:

Squared distance between rows: $$d^2(i,i') = \sum_{j=1}^{J} \frac{1}{f_{\cdot j}} \left(\frac{f_{ij}}{f_{i\cdot}} - \frac{f_{i'j}}{f_{i'\cdot}} \right)^2.$$

Squared distance between columns: $$d^2(j,j') = \sum_{i=1}^{I} \frac{1}{f_{i\cdot}} \left(\frac{f_{ij}}{f_{\cdot j}} - \frac{f_{ij'}}{f_{\cdot j'}} \right)^2.$$

A quadratic form is required so as to take advantage of multidimensional Euclidean geometry, without which computations would be tedious and results would lose certain properties. The distances defined above are known as χ^2 distances.

In our example, no word pairs or document pairs have identical profiles, so all distances are non-zero. Nevertheless, the words *family* and *health* have very similar profiles, and are used respectively 705 and 612 times. The occurrences of *family* are distributed among the six document categories as follows, rounded to the nearest integer to simplify comparisons: (21%, 17%, 6%, 27%, 23%, 6%) (see Table 2.3). The distribution of occurrences of *health* among documents is very similar, equal to (23%, 16%, 5%, 29%, 23%, 4%). The largest difference, which slightly exceeds 2%, is observed in the case of *women with low education*.

2.3.3 Inertia of a cloud

The *inertia* of a cloud is the sum of the inertia of each of its points, where the inertia of a point is defined as the product of its weight and its squared distance to the cloud's CoG. The total inertia, which for a given set of weights increases as distances do, is a measure of a cloud's dispersion. If all profiles are equal to one another, all points merge with the CoG, and the cloud's dispersion and inertia are both equal to zero.

When the reference axes are orthogonal, which is the case for both N_I and N_J, a cloud's inertia is simply the sum of the inertias of all the axes. On a given axis, the inertia of a point with respect to the CoG is equal to the product between its weight and its coordinate value squared on that axis. The inertia of the cloud N_I with respect to its CoG G_I is equal to:

$$I(N_I) = \sum_{i=1}^{I} f_{i\cdot} \sum_{j=1}^{J} \frac{1}{f_{\cdot j}} \left(\frac{f_{ij}}{f_{i\cdot}} - f_{\cdot j} \right)^2 = \sum_{i=1}^{I} \sum_{j=1}^{J} \frac{(f_{ij} - f_{i\cdot} \times f_{\cdot j})^2}{f_{i\cdot} \times f_{\cdot j}} = \frac{\chi^2}{N} = \Phi^2.$$

The *contribution* of each document i to the total inertia of the cloud N_I is thus proportional to its contribution to the χ^2 statistic.

In the same way, the inertia of N_J with respect to its CoG G_J is equal to:

$$I(N_J) = \sum_{j=1}^{J} f_{\cdot j} \sum_{i=1}^{I} \frac{1}{f_{i\cdot}} \left(\frac{f_{ij}}{f_{\cdot j}} - f_{i\cdot} \right)^2 = \sum_{i=1}^{I} \sum_{j=1}^{J} \frac{(f_{ij} - f_{i\cdot} \times f_{\cdot j})^2}{f_{i\cdot} \times f_{\cdot j}} = \frac{\chi^2}{N} = \Phi^2.$$

The *contribution* of each word j to the total inertia of the cloud N_J is proportional to its contribution to the χ^2 statistic.

The inertia of both clouds has the same value, denoted Φ^2, equal to the χ^2 value divided by N, the grand total of the lexical table.

Whereas the χ^2 test assesses independence between the documents and the words, the inertia Φ^2, independent of the grand total N of the lexical table, measures the strength of the relationship between them, which can vary greatly in textual data, depending of the specific nature of particular documents.

In our example, the inertia of both clouds (Φ^2=0.054) is low because many of the same topics (and thus words) are mentioned by different categories when asked about important things in life. This is actually a positive result in that it means people in different categories have similar concerns and can share their views, despite differences in age, gender, and education.

2.4 Fitting document and word clouds

Both the document and the word clouds can be visualized. Their shapes, intrinsically linked to the distance that is used, can be very informative. In our example, with six documents and ten words, each cloud sits in a 5-dimensional space; the dimension of the space is equal to the smallest dimension of the table (here: 6), minus 1, because the sum of the coordinates of the rows/columns is always equal to one (= sum of the profile's elements). Although this is a small example, both clouds of profiles (rows and columns) are not easy to visualize. Instead, we need to project them onto a sequence of planes, beginning with the one which retains the largest possible amount of inertia.

2.4.1 Factorial axes

Rows (documents) and columns (words) are represented by their profiles. Both clouds of row and column profiles are then centered, as the interest is in the dispersion of the clouds around their CoGs. In each space, the sequence of orthogonal *axes of maximal inertia*) (also called *factorial axes*), passing through the CoG, is then calculated.

In the row space, the factorial axes are denoted by u_s. The first axis u_1 maximizes the inertia of the projected cloud. Each subsequent axis u_s, orthogonal to the $(s-1)$ previous ones, maximizes the projected residual inertia. Analogously, in the column space, the factorial axes are denoted by v_s. The first axis v_1 maximizes the inertia of the projected cloud. Each subsequent axis v_s, orthogonal to the $(s-1)$ previous ones, successively maximizes the projected residual inertia.

It has been established that computing the factorial axes is equivalent to a certain matrix diagonalization, with the unit direction vectors u_s corresponding to the eigenvectors of this matrix. The first factorial axis corresponds to the eigenvector associated with the largest eigenvalue, denoted λ_1. The inertia along this axis is equal to λ_1. The second axis corresponds to the eigenvector associated with the second eigenvalue, λ_2, and so on. There are as many factorial axes as there are non-zero eigenvalues. Similarly, the unit direction vectors v_s corresponding to the axes of maximal inertia of the cloud of column profiles are the eigenvectors of the transpose of the above matrix, which has the same non-zero eigenvalues.

This implies that the column profile and row profile clouds, projected on to axes of the same rank s, have the same inertia, equal to λ_s. Both clouds sit in spaces with the same number of dimensions and have the same number of factorial axes with non-zero inertia, which is at most equal to $S = \min(I - 1, J - 1)$.

Indeed, the rows and columns come from the same table and thus correspond to two sides of the same data. The two clouds are strongly related, and this relationship manifests itself in various ways, here in terms of the two clouds having the same inertia, and later in terms of the *transition formulas* between row and column coordinates (see Section 2.4.2.3). These connections are referred to as *duality relations*. Duality reflects the two-sided approach we take to LTs: document-based and word-based.

In either space, looking for the plane of maximal inertia, or instead the first axis of maximal inertia followed by the second, leads to the same solution. More generally, the s-dimensional hyperplane of maximal inertia contains the corresponding $s-1$ dimensional one, and so on down to the plane and then axis of maximal inertia. That is, the subspaces of maximal inertia, known as *factorial subspaces*, are nested.

In CA, eigenvalues are non-negative and less than or equal to 1. An eigenvalue of 1 corresponds to a perfect association between a subset of rows (one row even) and a subset of columns (idem). This occurs when a subset of documents exclusively uses certain words which are never used in the other documents. In such cases, by reordering rows and columns of the lexical table, two disjoint subtables appear. $S' \leq S$ eigenvalues equal to 1 can be observed in the case of $(S' + 1)$ subsets of documents exclusively associated with $(S'$

+ 1) subsets of words. By reordering rows and columns of the lexical table, $(S' + 1)$ disjoint subtables would then appear. This situation is rarely seen however.

Nevertheless, in practice, large first eigenvalues (over 0.5, to give an order of magnitude) are frequently observed when CA is applied to short documents such as free text answers; in general, this occurs when a small subset of the documents uses a specific vocabulary and shares very few words with the other documents. In such cases, the first CA axis contrasts these documents (in the word space, specific vocabulary) with all of the other documents (in the word space, all other words). This axis therefore highlights a particular situation, not a general trend, and is therefore of little interest. When this happens, such documents should instead be defined as supplementary rows.

The total inertia of the cloud N_I (resp. N_J) is the sum of the inertias along each axis, i.e., the sum of the eigenvalues. The inertia associated with a factorial plane is equal to the sum of the two eigenvalues associated with its two axes. Consequently, the *percentage of inertia* that is *explained* by a given axis is equal to the ratio of its eigenvalue with respect to the sum of all the eigenvalues.

Given an $(I \times J)$ lexical table, the maximum value of Φ^2 is equal to $\min(I - 1, J - 1)$. This maximum is attained when the non-zero eigenvalues are all equal to 1. Cramér's V relates the observed value of the inertia (denoted by Φ^2) to its theoretical maximum value via the following ratio:

$$V = \left(\frac{\Phi^2}{\min\{(I-1),(J-1)\}} \right)^{\frac{1}{2}}.$$

This statistic varies between 0 (independence between documents and vocabulary) and 1 (maximum association between documents and vocabulary). Assuming that the number of documents is less than the number of words $(I < J)$, this maximum is reached when each document uses its own set of words, none of which are found in the other documents. Cramér's V allows us to compare the inertia of tables of different sizes.

In our example, the total inertia (equal to 0.054) can be decomposed, axis by axis, as shown in Table 2.6 and Figure 2.1. This inertia and Cramér's V $(= (0.054/5)^{\frac{1}{2}} = 0.10)$ are both weak. Nevertheless, a low total inertia does not forbid the existence of certain factorial axes which uncover interesting structure in the data.

2.4.2 Visualizing rows and columns

Let F_s and G_s denote, respectively, the vectors of row and column coordinate values projected on the axis of rank s. We refer to these as *factors* of rank s on, respectively, the rows and columns. These can be considered as quantitative variables on the rows and columns. The variances of factors of the same rank s are equal—and in fact equal to λ_s. These coordinate values map the row points

TABLE 2.6
CA applied to the running example. Excerpt of the summary output.

```
Correspondence analysis summary

Eigenvalues
      Variance % of var. Cumulative % of var.
dim 1    0.032    59.182                59.182
dim 2    0.016    30.487                89.669
dim 3    0.003     5.844                95.512
dim 4    0.002     3.404                98.916
dim 5    0.001     1.084               100.000

Cramer's V  0.104    Inertia  0.054


DOCUMENTS
All documents are aggregate documents

 Coordinates of the documents
          Dim 1  Dim 2  Dim 3  Dim 4  Dim 5
M_EduLow  -0.058 -0.224  0.022 -0.008  0.018
M_EduMed   0.194 -0.021 -0.011  0.067 -0.018
...

 Contribution of the documents (by column total=100)
          Dim 1  Dim 2  Dim 3  Dim 4  Dim 5
M_EduLow   2.132 62.468  3.120  0.768 11.124
M_EduMed  22.329  0.528  0.706 46.759 10.825
...

 Squared cosinus of the documents (by row total=1)
         Dim 1 Dim 2 Dim 3 Dim 4 Dim 5
M_EduLow  0.061 0.923 0.009 0.001 0.006
M_EduMed  0.874 0.011 0.003 0.105 0.008
...


WORDS

 Coordinates of the words
          Dim 1  Dim 2  Dim 3  Dim 4  Dim 5
children  -0.280  0.183 -0.069 -0.049  0.045
family    -0.047  0.003  0.016 -0.016  0.014
...

 Contribution of the words (by-column total=100)
          Dim 1  Dim 2  Dim 3  Dim 4  Dim 5
children  12.983 10.776  7.901  6.894 18.536
family     1.935  0.014  2.319  3.849  9.540
...

 Squared cosinus of the words (by-row total=1)
         Dim 1 Dim 2 Dim 3 Dim 4 Dim 5
children  0.648 0.277 0.039 0.020 0.017
family    0.754 0.003 0.089 0.086 0.068
...
```

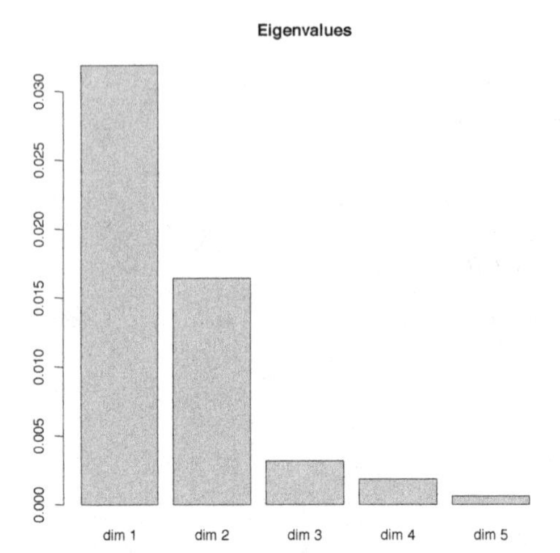

FIGURE 2.1
Eigenvalues associated with the factorial axes.

(resp. column points) on to orthogonal axes in classical Euclidean spaces. Thus, CA embeds the row and column spaces, endowed with χ^2 distances, into classical Euclidean spaces where the distances between points can be visually evaluated (see Figure 2.2).

2.4.2.1 Category representation

Figure 2.2 (top) shows that the categories are ranked by increasing the level of education, from left to right, on the first axis. In fact, the axis is not oriented, so it would be entirely equivalent to observe that the categories are arranged by decreasing level of education from left to right. What is noteworthy is that the order is recovered, even though it was unknown when computing the axes and coordinates. Therefore, lexical choices are markers of the level of education as well as other social characteristics such as gender—which is clearly associated with the second axis here, contrasting the male categories (bottom) with the female ones (top).

2.4.2.2 Word representation

Looking at the word map in Figure 2.2 (bottom), we see that the first axis contrasts—to the left—*children, health, family,* and *home* (= the family world) with—on the right—*work, friends,* and *job* (= the external world). As for the second axis, it contrasts *life* and *work* (negative coordinate values) with *friends* (positive coordinate values). To understand why, it is necessary to look closer at what are known as the *transition formulas* between the spaces.

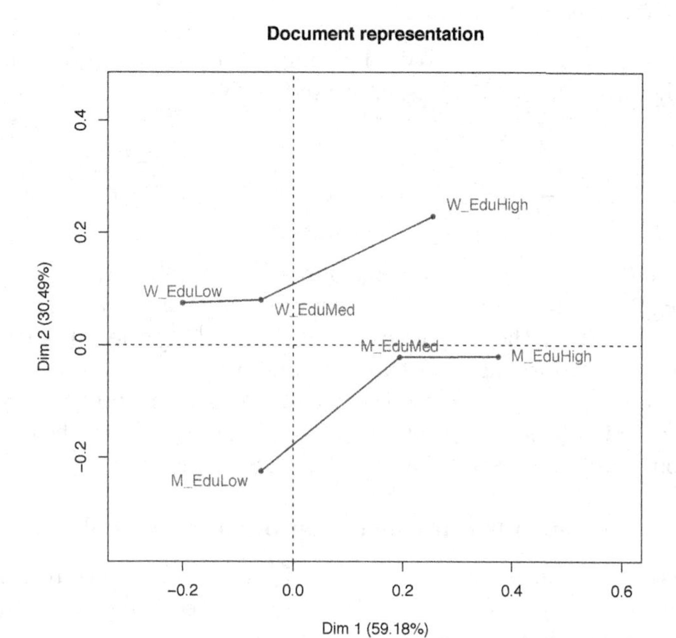

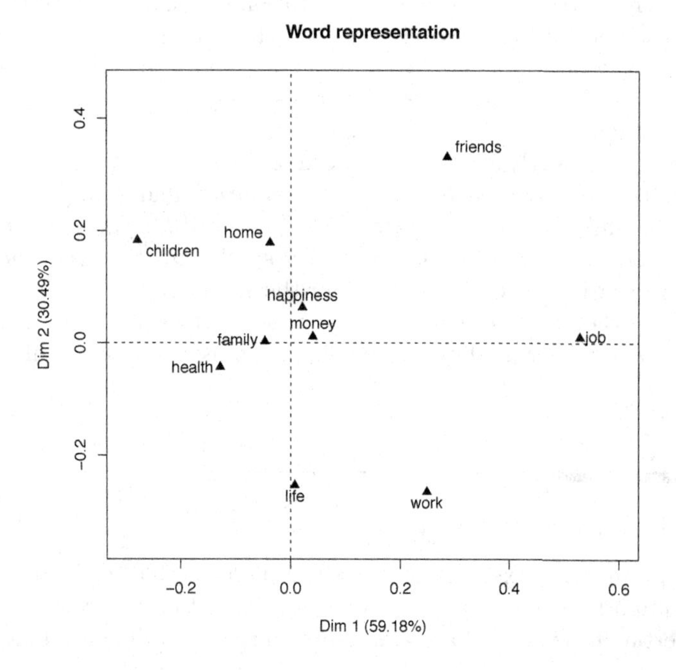

FIGURE 2.2
CA applied to the small example. Top: row representation; bottom: column representation.

2.4.2.3 Transition formulas

On axis s, the factors on the rows and columns, F_s and G_s, are connected by the so-called *transition formulas*, expressed as follows:

$$F_s(i) = \frac{1}{\sqrt{\lambda_s}} \sum_{j=1}^{J} \frac{f_{ij}}{f_{i\cdot}} G_s(j), \qquad G_s(j) = \frac{1}{\sqrt{\lambda_s}} \sum_{i=1}^{I} \frac{f_{ij}}{f_{\cdot j}} F_s(i) .$$

$F_s(i)$ denotes the coordinate value of row i on axis s, $G_s(j)$ the coordinate value of column j on axis s, λ_s the eigenvalue associated with axis s, $f_{i\cdot}$ the weight for row i, $f_{\cdot j}$ the weight for column j, and f_{ij} the proportion of the occurrences corresponding to document i and word j.

Therefore, on axis s, the row point i is, up to a constant, at the CoG of the column points j, and vice versa. The weights are given by the components of row point profile i (resp. column point profile j).

2.4.2.4 Simultaneous representation of rows and columns

The transition formulas allow us to move from the row space to the column space and vice versa, thus permitting the simultaneous representation of the rows and columns on the same graph, as shown in Figure 2.3. On these axes, words are attracted by categories which over-use them, and repelled by those which under-use them. However, interpretation of the simultaneous representation must obey the *barycentric principle*, i.e., the position of a row point depends on the positions of all column points, and vice versa. A document i and word j can be close together without actually attracting each other, due to their relationships with other points.

The first axis clearly contrasts the word *job*, over-used by the *high education* categories regardless of *gender*, to the words *health* and *family*, over-used by the *medium* and *low education* categories. The word *happiness* is close to the CoG as it is used by all categories with a similar relative frequency (so its profile is close to the average one). As for the second axis, while *life* and *work* stand out on the negative side due to over-use by the three categories of men, the word *friends*, over-used by women, has the most positive coordinate value.

2.5 Interpretation aids

The interpretation of axes and planes is based on standard real-valued indices, which is important in CA since the contributions of points (rows or columns) to the inertia of the axes depends not only on their coordinate values, but also their weights.

In textual CA, such indices are used to select the words and documents

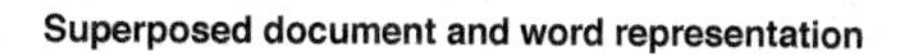

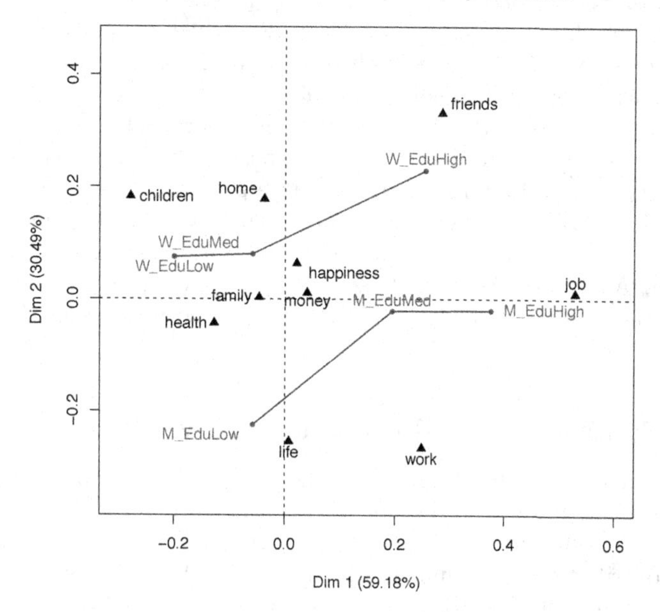

FIGURE 2.3
Simultaneous representation of documents and words.

on which interpretations have to be based, as well as those which need to be represented, and filter out the rest. This type of filtering is important since, in real applications, there can be a huge number of words and documents (typically, from a hundred to several thousand).

2.5.1 Eigenvalues and representation quality of the clouds

The total inertia of both clouds, equal to the sum of the eigenvalues, measures the strength of the relationship between documents and vocabulary. Each axis expresses part of this relationship, and its importance scales with the size of its associated eigenvalue. The percentage of inertia associated with an axis, plane, or higher-dimensional subspace corresponds to the representation quality of the cloud (rows or columns) when projected onto the corresponding subspace.

Usually, eigenvalues are visualized using a bar chart in which the height of the bar associated with each eigenvalue is proportional to its relative importance (see Figure 2.1). In decreasing order, this sequence of eigenvalues shows the relative importance of the axes, and helps us to decide how many axes or planes need to be interpreted. Essentially, we choose to interpret axes individually if their associated eigenvalue is clearly different to the previous and subsequent ones. If instead two eigenvalues are close to each other but well separated from all others, we will interpret the corresponding plane rather than each of them separately. A large jump in the sequence can help decide empirically which axes to examine further and which to put aside.

In our example, the total inertia is equal to 0.054. The first two axes clearly dominate, with eigenvalues equal to 0.032 (59.2% of the inertia) and 0.016 (30.5%), summing to 89.7%. Then, a large gap separates the second eigenvalue from the third. It is thus likely that the first factorial plane corresponds to the majority of useful information. The large percentage of inertia associated with the first plane permits us to put aside worries about small amounts of distortion in the distances between documents and between words due to this projection. As seen in Section 2.4.2.1, the first two axes each have a clear interpretation, which accords well with the gap seen between the first two eigenvalues.

2.5.2 Contribution of documents and words to axis inertia

The total inertia can be decomposed, document by document, or word by word. The highest-contributing documents and words provide relevant information about the structure of both the document and word clouds.

The contribution of document i to the inertia λ_s of axis s is equal to the inertia of its projection onto this axis with respect to the CoG. It is therefore computed as the squared coordinate value $F_s^2(i)$ of its projection, multiplied by its weight $f_{i\cdot}$. Generally, this contribution, denoted $ctr_s(i)$, is given as a percentage of the total inertia of the axis:

$$ctr_s(i) = 100 \times \frac{f_{i\cdot} \times F_s^2(i)}{\lambda_s}.$$

The sum of contributions of all documents to the inertia of an axis is therefore equal to 100% of the inertia of the axis, i.e.,

$$\sum_{i=1}^{I} ctr_s(i) = 100.$$

Similarly, the contribution of word j to the inertia of axis s is given by:

$$ctr_s(j) = 100 \times \frac{f_{\cdot j} \times G_s^2(j)}{\lambda_s},$$

and

$$\sum_{j=1}^{J} ctr_s(j) = 100.$$

Going back to our example, the documents contribute in a rather balanced way to the first axis's inertia, except for *men with low education* and *women with medium education*, whose contributions are less than 2.5% (the

`TextData.summary` function provides this information—see Table 2.6 for partial results).

This axis is therefore a *general axis*, expressing an overall trend in the data, here a gradual change in vocabulary according to education level. In contrast, a single document, *men with low education*, contributes 62.5% of the inertia of the second axis. Therefore, this is a *specific axis*, highlighting a precise piece of information. Indeed, the vocabulary in this category disproportionately involves the word *life* in expressions like *to enjoy life, enjoying life, healthy life, good family life*, and so on. Note that it is quite often the case that the first axes reflect general trends in the data, while later ones highlight more specific phenomena.

We should also not forget that the relatively small vocabulary in our example makes interpretations somewhat simplistic. However, it turns out that when the same data is used but the usual thresholds on word frequency are in place, similar though more stable structure appears, as we will see in the next chapter.

2.5.3 Representation quality of a point

As we said earlier, the percentage of (the total) inertia associated with a given axis is a measure of the overall representation quality of the cloud as projected on that axis. To evaluate the *representation quality* of a point (row or column) on axis s, the same principle can be applied: the ratio of the inertia of the point i projected on to axis s with respect to its inertia in the full space, denoted $qlt_s(i)$, is computed as:

$$qlt_s(i) = \frac{\text{inertia of point } i \text{ projected on axis } s}{\text{total inertia of point } i} = \frac{f_{i\cdot}F_s^2(i)}{f_{i\cdot}d^2\left(i, G_I\right)} = \frac{F_s^2(i)}{d^2\left(i, G_I\right)}.$$

The representation quality of a row point is also known as the *relative contribution* or *cos*^{2}. The sum of the representation qualities of a point on the successive axes s ($s = 1, \ldots, S'$) measures the representation quality of the point on the associated S'-dimensional subspace. If $S' = S$, then the point is perfectly represented because:

$$\sum_{s=1}^{S} qlt_s(i) = 1.$$

Similarly, the representation quality of the column point j on the axis s is computed as:

$$qlt_s(j) = \frac{\text{inertia of point } j \text{ projected on axis } s}{\text{total inertia of point } j} = \frac{f_{\cdot j} \cdot G_s^2(j)}{f_{\cdot j} \cdot d^2\left(j, G_J\right)} = \frac{G_s^2(j)}{d^2\left(j, G_J\right)},$$

and

$$\sum_{s=1}^{S} qlt_s(j) = 1.$$

The representation quality of a row or column point on axis s is denoted `cos2` in the CA output since it is also equal to the squared cosine between the vector connecting the point to the CoG, and its projection on to axis s.

Only two documents—those which did not contribute to the first axis (*men with low education* and *women with medium education*)—are poorly represented by the first axis (with respectively $\cos^2 = 0.1$ and 0.2). The former, *men with low education*, is well represented by the second axis ($\cos^2 = 0.9$) and it is the only such one.

2.6 Supplementary rows and columns

2.6.1 Supplementary tables

Supplementary (or *illustrative*) rows and/or columns can be projected onto the factorial axes or planes. These come from *supplementary tables* involving either active documents versus supplementary quantitative, qualitative, or frequency columns, or active words versus new documents (called *supplementary documents*). Supplementary points (rows or columns) do not need to constitute homogeneous sets since computing each of their positions is carried out independently.

Usually, contextual variables (either quantitative or qualitative) describing the documents are projected as supplementary columns. Repeated segments can also be projected as supplementary frequency columns to better capture the meaning of words, which thus appear in their proper context. Regarding the row space, a subset of documents—such as the male categories for instance—could be projected as supplementary rows on the factorial axes computed from only the female categories.

Thus, factorial planes can be enriched with external information that was not involved in computing the axes; this can play a paramount role in the interpretation step.

2.6.2 Supplementary frequency rows and columns

By using the transition formulas, the coordinate value of supplementary row i^+ on axis s is computed using the vector of coordinates of the columns G_s, the eigenvalue λ_s, and the profile $(f_{i^+j}/f_{i^+\cdot})$ of this row i^+, as follows:

$$F_s(i^+) = \frac{1}{\sqrt{\lambda_s}} \sum_{j=1}^{J} \frac{f_{i^+j}}{f_{i^+.}} G_s(j).$$

Analogously, the coordinate value of supplementary column j^+ on axis s is computed using the vector of coordinates of the rows F_s, the eigenvalue λ_s, and the profile $(f_{ij^+}/f_{.j^+})$ of this column j^+, as follows:

$$G_s(j^+) = \frac{1}{\sqrt{\lambda_s}} \sum_{i=1}^{I} \frac{f_{ij^+}}{f_{.j^+}} F_s(i).$$

2.6.3 Supplementary quantitative and qualitative variables

It is necessary to take into account that the contextual variables are defined at source document level. The **Xplortext** package allows us to visualize the contextual variables, as detailed below.

In a direct analysis, the coordinate value of a quantitative contextual variable on factorial axis s is equal to the correlation coefficient between this variable and the factor on the documents of rank s (= the vector of coordinates of the documents on the rank s axis). Qualitative variables are displayed in terms of their categories, mapped on each axis to the CoG of the documents that belong to them. When dealing with an aggregate analysis, on the other hand, the contextual variables have to be recoded. In the case of a quantitative variable, the average by category is computed in order to obtain a quantitative variable defined at the aggregate document level. Then, on factorial axis s, the coordinate value of this variable is computed as its correlation coefficient with the factor on the documents of rank s. As for a qualitative variable, new aggregate documents are formed consistent with its categories and then projected as supplementary documents (= supplementary rows) on to the axes.

2.7 Validating the visualization

The representations of words and documents need to be validated by drawing *confidence ellipses* around them. The **Xplortext** package provides replicate samples as follows. The $I \times J$ cell counts of replicate lexical tables are considered to follow a multinomial distribution with as many categories as there are cells in the table. The $I \times J$ theoretical probabilities of success are estimated by supposing they are equal to the proportions computed when transforming the original lexical table into a table of proportions.

Successively, many replicate samples of the same size as the original one are randomly drawn from this distribution and the associated replicate lexical tables built. Then, the row documents/column words of all replicate tables are projected as supplementary rows/columns onto the CA planes computed from the original lexical table. The cloud of replicate points corresponding to the same document/word is represented by an ellipse drawn in such a way that the most extreme $\alpha\%$ points are excluded when a confidence level equal to $(1-\alpha)\%$ is chosen. Usually, representing all of the confidence ellipses leads to an unreadable graph. It is therefore necessary to select the points around which ellipses are drawn. Different criteria can be used, such as the most highly-contributing documents and words, or those with good representation quality. Certain words/documents can also be selected based on their relevance in the interpretation step.

In the category case, overlapping confidence ellipses means that the corresponding documents do not use clearly different vocabulary. Figure 2.4 shows that this occurs for the categories *women with high education*, *men with high education*, and *men with medium education*, found to the right on the first axis. The same occurs for the categories *women with medium education* and *women with low education*. On the contrary, *men with low education* have a specific vocabulary. This allows us to refine the interpretation given in Section 2.4.2.4. The first axis contrasts *men and women with high education* and *men with medium education* with the other categories. The second axis highlights a difference between genders, limited to the less educated categories; we see that *men with medium education* are close to *men with high education*, while *women with medium education* are close to *women with low education*. We should try to investigate the reasons why, which would certainly require more than the current data. As frequently, this first exploratory analysis suggests new hypotheses which can lead to complementary studies.

In the words case, a confidence ellipse overlapping the CoG indicates that the word is used in the same way by all documents. In the case of two words that are thought to be potentially synonymous, overlapping confidence ellipses mean that they are synonyms in the context of the corpus. In our example, we can deduce for instance that *job* and *work*, whose ellipses only slightly overlap, are relatively different words in the framework of this corpus.

2.8 Interpretation scheme for textual CA results

In earlier sections, we have described the principles of CA. This information can then be used in the interpretation process as follows:

- **Bar chart of the eigenvalues**. The sequence of eigenvalues and rates of inertia provide information on the relative importance of the axes. Big jumps attract attention, but are not always decisive in choosing which axes should be retained, since their interpretability is of greater importance. Interest in an axis stems from a sense of the associations it uncovers. In

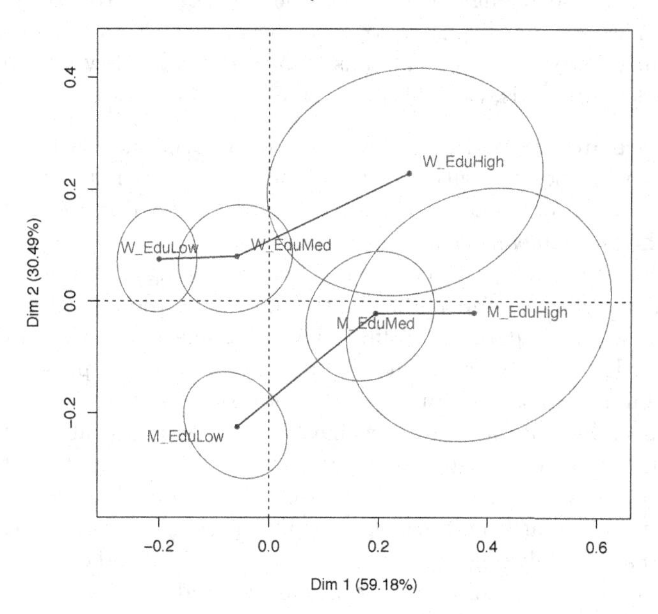

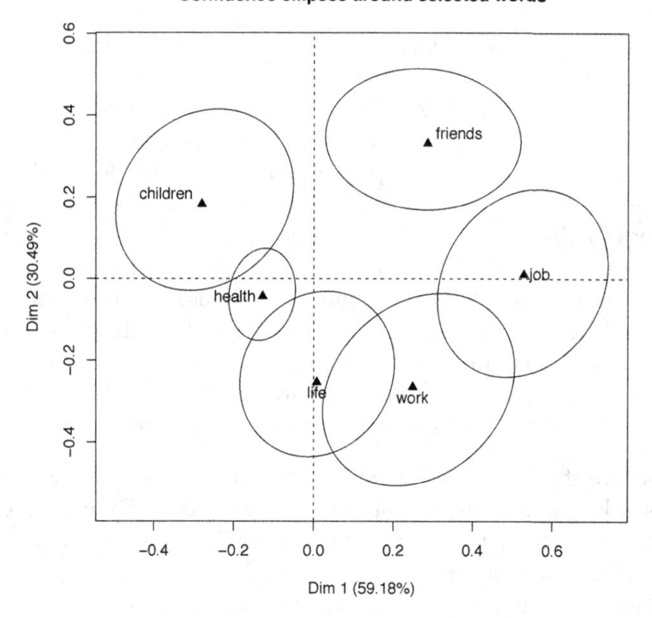

FIGURE 2.4
First CA plane. Validation of the representations. Top: representation of the confidence ellipses around documents; bottom: representation of the confidence ellipses around selected words; only those which do not cover the CoG are represented.

textual CA, the inertia (= sum of the eigenvalues) is considered to be a measure of the strength of the relationship between row documents and column words. Each axis contributes according to its percentage of the total inertia. Nevertheless, an axis associated with a low percentage, even if it is the first axis, can still provide useful information.

- **Interpreting axes**. Interpretation of a factorial axis is based on high-contributing points (without which the axis would not exist) as well as points having both large coordinate values and good representation quality. The latter means those whose difference to the average profile is large and almost entirely characterized by the axis being interpreted. It is usual practice to characterize axes using the row points (documents) and column points (words) separately at first. Then, associations between documents and words should be taken into account. Points whose representation quality is high are also relevant for interpretation. To uncover the meaning of an axis and put it into words, we have to understand what it is that points with negative coordinates share, and what it is that points with positive coordinates share, at least in the case of high-contributing points, which are not always far from the CoG. The interpretation of an axis can later be refined after analyzing subsequent axes. If an initial and overly quick interpretation leads to finding similar proximities and contrasts on two different axes, a more in-depth search will be necessary to find out what differentiates them. Inspecting the plane spanned by these two not-necessarily contiguous axes could help with this.

- **Interpreting planes**. Once the axes have been characterized, an inspection of certain factorial planes can help to summarize and refine interpretations. Further features to those initially seen in individual axes may appear. In CA, it is legitimate and common to represent rows and columns on the same graph. This graph helps us visualize the words among the documents, and the documents among the words. Nevertheless, the high number of points (especially of word points) in textual CA often pushes us to adopt separate representations of the two sets which need to be examined simultaneously, keeping in mind the transition formulas. The shapes of the point clouds are important information. Usually, either clearly separated clusters of documents or words can be seen, or clouds with triangular or parabolic shapes. Examples presented later in the book will show how the shape—a structural feature—can be interpreted.

- **Supplementary elements**. Supplementary elements (rows or columns) play a relevant role in the interpretation of axes. By construction, they contribute nothing to the inertia, but their representation quality is a measure of their relevance. Thus, for each axis or plane, well-represented supplementary elements can also be taken into account.

- **Returning to the data**. It can often be useful to build on discovered proximities, contrasts, and other observed structure—such as trajectories seen on axes and planes—by going back to the profile tables, looking for certain profiles or elements. This can help place structural features in their initial context, allowing us to better understand how elements that stand out differ from the others.

2.9 Implementation with Xplortext

Our script file `script.chap2.R` calls the `Aspiration_Int_UK.RData` database and performs the following tasks:

- The `TextData` function selects and encodes the corpus as an ALT, in this case to obtain a reduced-size ALT.
- Various R commands allow us to visualize the tables used in CA that are shown throughout the chapter.
- The `LexCA` function applies CA to the ALT.
- The `plot.LexCA` function provides several CA graphs.
- The `ellipseLexCA` function gives, successively, the representations of documents and active words, enriched by confidence ellipses.

2.10 Summary of the CA approach

CA provides a simplified representation of lexical tables, helping to visualize proximity between documents and between words, and associations between documents and words, on a small number of factorial axes and planes, with nothing essential being lost. The set of documents is then displayed in terms of the distances between their lexical profiles, and the set of words by the distances between their relative distributions among documents. The similarities and differences between documents can be matched up with those between words that explain them, and vice versa. Keep in mind that words take their meaning from their lexical environment. CA starts precisely from co-occurrences, and extracts structure from these. Here are the main points to remember:

- A distance between profiles is defined in both the word space and document space. These distances respect the principle of distributional equivalence. Thus, merging two documents with the same profile into a single document, or two words with the same profile as a single word, does not modify other distances. The distance we work with is called the χ^2 distance.

- The dispersion of both document and word clouds is measured in terms of their inertia. The total inertia of both clouds is the same.
- The main point of CA is to visualize these clouds on the axes of maximal inertia so as to unveil structure in the data, rather than focusing on dimension reduction.
- In the two clouds, axes of the same rank characterize the same part of the inertia.
- Document and word clouds are projected onto factorial axes. The coordinate vectors of documents/words, called factors, are computed by CA.
- Words frequently used in the same documents in the same way will be close together on factorial axes and planes.
- Words frequently used in the same context(s), even if never used in the same documents, will be close to each other on the factorial axes and planes. Such words retain semantic proximity, which allows us to detect synonyms.
- The transition formulas enable us to move from one space to the other. Up to a constant, documents are at the CoG of the words that they use, and words are at the CoG of documents that use them.
- The simultaneous representation of rows and columns, based on the transition formulas, enables us to link groups of words with groups of documents, thus shining a light on the content of documents. Given that the position of a document depends on all of the words' positions (and vice versa), proximity between a given document and a given word can be misleading. Consulting the association rates between documents and words can help to specify those which truly are related.
- CA maps the documents and words into spaces endowed with the classical Euclidean distance. Consequently, proximity can be detected visually.

Being able to interpret results is essential, and requires the kind of experience gained by analyzing many examples. In this context, the "Les Cahiers de l'Analyse des Données" journal is today still the largest collection of work relative to textual data analysis using CA. This journal has been digitized and is freely accessible at http://www.numdam.org/journals/CAD.

3

Applications of correspondence analysis

3.1 Choosing the level of detail for analyses

As discussed in Chapter 1, textual CA can be performed on the source documents or on the aggregate ones, starting respectively from an LT or an ALT.

In the CA setting, the differences between these two types of analysis derive mainly from the fact that the source documents are simultaneously more homogeneous and more different to one another than the aggregate documents are, leading to higher cloud inertia. As examples, we will examine the inertia calculated in the analysis of free text answers aggregated according to categories of the `Age_Educ` variable (see Section 3.2 of the current chapter), the direct analysis of these same free text answers (see Section 4.11) with a view to further clustering, and lastly, the direct analysis of 11 political speeches with around 10,000 occurrences each (see Section 7.4). In the second and third examples, each document, either an answer or a speech, comes from the same author, conferring it a high level of homogeneity and clearly distinguishing it from the other documents. Therefore, rather high inertia is expected. In the first example, the aggregate documents are composite, making them relatively heterogeneous, leading to a lower observed inertia. Given the different dimensions of the tables in the three examples, Cramér's V can be used for comparisons. The values of this turn out to be, respectively, 0.14, 0.44, and 0.27. We see indeed that the highest values are for the direct analyses (second and third examples).

Another important point concerns the contextual variables, defined at the source document level. The solutions adopted in the **Xplortext** package were discussed in Section 2.6.3. We will look at several examples in the following sections.

3.2 Correspondence analysis on aggregate free text answers

3.2.1 Data and objectives

We now return to the corpus analyzed in Chapter 2, this time selecting the words using more typical frequency thresholds. Here too, the free text answers are aggregated according to the six categories of the `Gender_Educ` variable.

We aim to study variability in primary vital concerns, as related to gender and education.

3.2.2 Word selection

Words are selected if used in at least 15 documents, i.e., by approximately 1.5% of the documents, consistent with the usual criterion of keeping words used by at least 1–2% of users. Here, stopwords are meaningful (in particular, the personal pronouns) and thus not suppressed. These choices lead to 135 of the 1334 words being kept, as well as 78% of occurrences, i.e., 10,818 of the initial 13,917 (see Table 3.1). A high level of stability in the CA results is seen when we vary the frequency threshold between 10 and 20 documents.

3.2.3 CA on the aggregate table

CA can then be applied to the *Gender_Educ* $\times$ *words* table. The words *wife* and *husband* are considered as supplementary words because of being used almost exclusively by either men or women. If active, they would have had too much influence on computing the first axes. The main results are shown in Figure 3.1 and Table 3.2.

The values of the inertia (= 0.10) and Cramér's V (= 0.14) are low but typical for this kind of analysis. In line with the low total inertia, the eigenvalues are small, as is of course Cramér's V. We are therefore far from the largest possible inertia, equal to 5 in this case, which would be seen if each of the categories had its own unique vocabulary. As for the smallest possible inertia, equal to 0, it corresponds to the case where the lexical profiles are equal to each other and equal to the average (i.e., marginal) profile.

On the other hand, the percentages of inertia corresponding to the first axes are high. The first axis dominates (37.9% of the inertia) and the principal plane retains 60.1% of the inertia, which is usual in this type of analysis.

These findings imply the existence of structure in the data involving preferential relationships between certain words and document categories, thus justifying the utility of our analyses. Some categories emphasize some aspect of life or another.

However, the inertia is low because we find ourselves in the common situation whereby respondents speak the same language (and use many of the

TABLE 3.1
Excerpt from the CA results: corpus size, vocabulary size, and index of the most frequent words.

```
TextData summary
                         Before After
Documents               1043.00     6
Occurrences            13917.00 10818
Words                   1334.00   135
Mean-length               13.34  1803
NonEmpty.Docs           1040.00     6
NonEmpty.Mean-length      13.38  1803

Index of the  20  most frequent words
        Word Frequency N.Documents
1  my              810           6
2  family          705           6
3  health          612           6
4  to              523           6
5  and             504           6
6  the             332           6
7  of              312           6
8  good            303           6
9  a               300           6
10 i               287           6
11 happiness       229           6
12 in              181           6
13 money           172           6
14 life            161           6
15 that            160           6
16 job             143           6
17 is              141           6
18 happy           137           6
19 be              136           6
20 children        131           6
```

same words). Irrespective of age and gender, many of them even share some of the same opinions, which is somewhat reassuring! The lexical differences between the documents are small, and a solid interpretation of the results will require us to frequently return to the profile tables and/or the association rate table of documents × words, which is easier to read.

The large jump between the values of the second and third eigenvalues suggests limiting the study to the first two axes. All the same, we will take a brief look at the third one before choosing whether to discard it definitively.

3.2.3.1 Document representation

On the CA's first factorial axis the *a priori* order induced by level of education is clearly visible, both for the male and female categories (see Figure 3.1). The

TABLE 3.2
Excerpt from the CA results: first five eigenvalues, Cramér's V, and total inertia.

```
Correspondence analysis summary

Eigenvalues
      Variance % of var. Cumulative % of var.
dim 1    0.039    37.938               37.938
dim 2    0.023    22.197               60.135
dim 3    0.016    15.183               75.318
dim 4    0.014    13.833               89.151
dim 5    0.011    10.849              100.000

Cramer's V  0.143    Inertia  0.102
```

evolution of concerns as the level of education changes, as conveyed by words, is a dominant structural feature. For both genders, low levels of education are set apart from the medium and high ones, which are themselves very close to one another. However, their vocabularies are sufficiently different that we obtain non-overlapping confidence ellipses. The second axis separates the genders; the two trajectories of the three levels of education are approximately parallel, although male and females with low levels of education are further from each other than the corresponding higher-educated respondents. Therefore, education and gender are related to lexical choices, and their respective influence is measured by the inertia associated with the two axes.

3.2.3.2 Word representation

First, we consider the representation of all retained words (see Figure 3.1). This cloud has a commonly observed shape: elongated along the first axis and denser in its central part.

Next, the plots of the highest-contributing and best-represented words (see Figure 3.2) on either of the first two axes are shown. Recall that the former are responsible for the construction of the axes; if deleted, the axes they contribute to would disappear. The latter are the most useful words for describing the axes; their profiles strongly differ from the average one, and this difference is almost entirely expressed by the axes on which they are best-represented. The selection criteria are defined in such a way that a maximum number of words can be plotted while maintaining legibility.

3.2.3.3 Simultaneous interpretation of the plots

As we saw before, the first axis in Figure 3.1 (top) shows the education categories in their natural order. Taking into account the transition formulas, this axis also contrasts words preferred by low-education categories—on the right, with those preferred by the medium- and high-education categories—on the

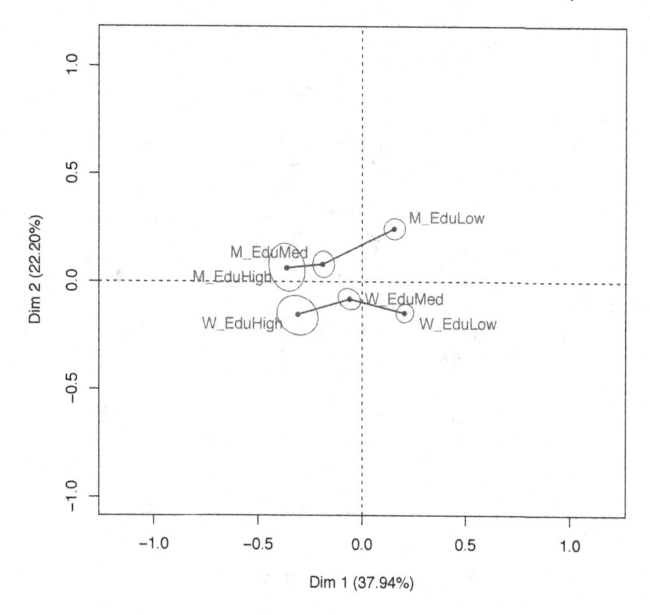

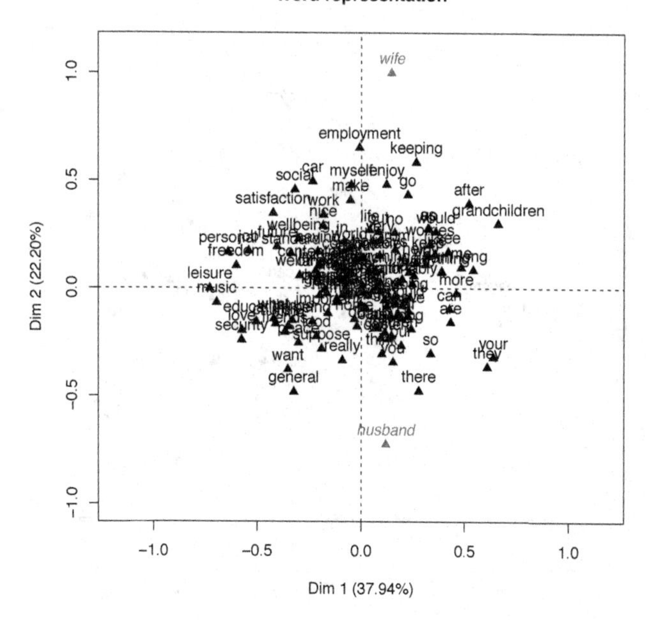

FIGURE 3.1
First CA plane. Top: representation of the documents and their confidence ellipses; bottom: representation of the words; supplementary words are in italics.

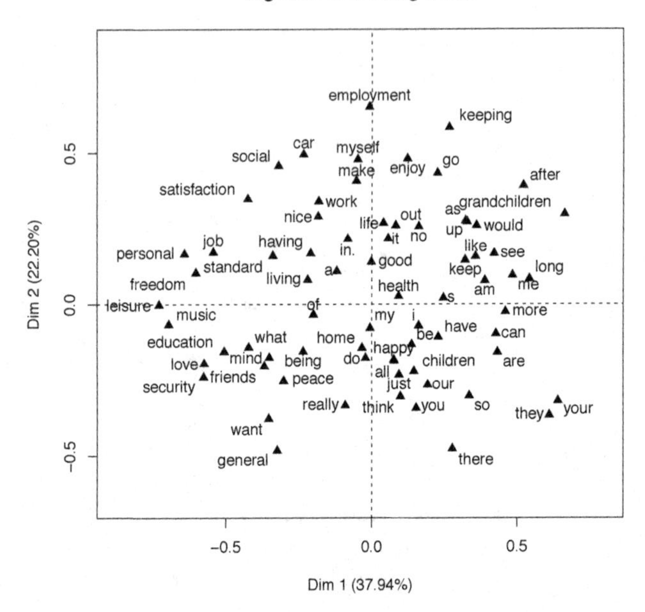

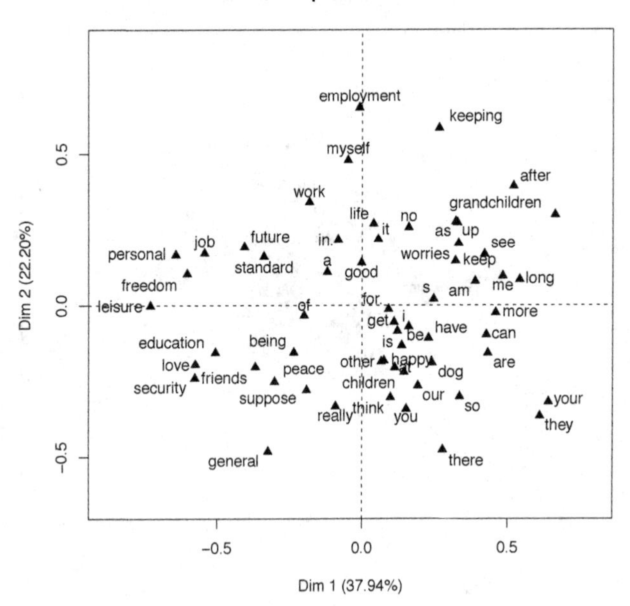

FIGURE 3.2
First CA plane. Top: highest-contributing words; bottom: words whose representation quality is greater than 0.60.

left. The vocabulary of the low-education categories does not correspond to some clear opinion; very few nouns are seen to the right, and indeed, function words are the most common, which indicates some difficulty or reluctance to express opinions. On the contrary, the two higher-education categories mainly use nouns related to the professional domain: *employment, work, job, education*, and personal activities: *leisure, music, love, friends*, as well as general ideas like *freedom, peace* and *security.*

The second axis contrasts the two genders, then the words respectively favored by each. Interpreting this in terms of gender, but also taking into account the level of education trajectories, may be useful. On the right-hand side of the first axis (= low-education), men favor *grandchildren* and basic verbs, while women over-use *children* and various pronouns. Then, with higher levels of education, the professional domain appears, mostly concerning men. This topic appears in the use of the words *employment, work*, and *job*, depending on the level of education. The positions of *job* and *employment* on the first plane suggest that both are *masculine words* in the sense that they are used relatively more often by men than by women. In this case, the frequencies can be directly compared as the counts of men and women are similar. As for the women's trajectory, when level of education increases, topics relevant to personal life appear, such as *friends, love*, and *security.*

3.2.4 Supplementary elements

Introducing supplementary rows and/or columns can enrich the results. These elements do not intervene in the construction of the axes, and are projected *a posteriori* onto them using the transition formulas (see Section 2.4.2.3).

3.2.4.1 Supplementary words

Here the two words *wife* and *husband* have been used as supplementary frequency columns. CA shows that they are almost exclusively used by men or women, given their extreme and opposing positions on the second axis. The method takes into account this almost-exclusive usage (though supplementary, these words are found at the most extremal positions on the second axis), but does not take into account the different frequency of usage of each—and this, irrelevant of whether they are in fact active or supplementary. This situation forces us to frequently return to the data, i.e., continually refer back to the lexical table being analyzed.

3.2.4.2 Supplementary repeated segments

A set of interesting and well-represented repeated segments are shown in Figure 3.3. These provide word context and thus meaning. For example, we can see that the respondents wish for a *good job* and/or *job satisfaction.* These characteristics are not associated with *work* or *employment*, though they could *a priori* have been considered as synonymous with *job*. We see also that *peace* is either *peace of mind* or *peace in the world*, clearly two different things.

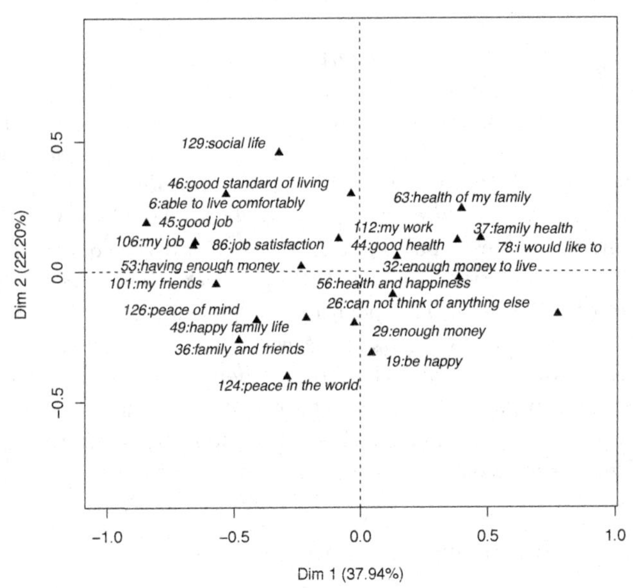

FIGURE 3.3
Repeated segments as supplementary columns.

3.2.4.3 Supplementary categories

We might wonder whether the education effect is spurious, as young people are generally more educated than seniors. To examine this further, we now want to consider the 18 categories of the `Gen_Age_Edu` variable as supplementary row documents. Note that, among them, three categories aggregate less than ten respondents: *men* and *women 30 years and under*, and *women over 55*.

For that purpose, the variable `Gen_Age_Edu` has to be introduced as a supplementary qualitative variable in the former analysis, using the argument `context.quali` of the function `TextData` (see Section 3.2.5). Then, this same function automatically manages the creation of the aggregated documents corresponding to the `Gen_Age_Edu` categories. Finally, the function `LexCA` will treat these document categories as supplementary rows. Since the active elements (rows and columns) have not been modified, the same eigenvalues and the same representation of the active words and row categories are obtained. Thus, the former representation of the documents is now enriched by that of the supplementary categories (see Figure 3.4). The six trajectories linking the same-age and same-gender categories show how vocabulary evolves with education, for fixed gender and age.

This example shows both the interest and point of supplementary elements. Furthermore, it illustrates how we can study two different segmentations of the corpus, with the more detailed one nested inside the other, in a way that leads to new understanding. We could also swap the roles of the active and supplementary rows, but then the first plane would not be so meaningful since the important information would be scattered across a greater number of axes.

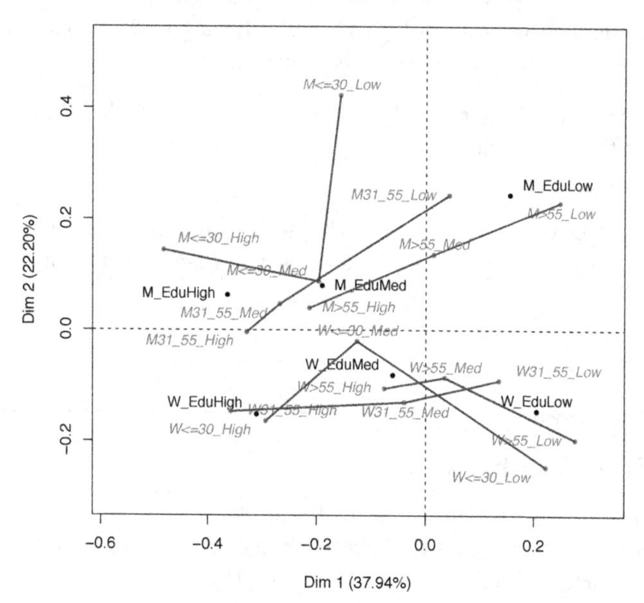

FIGURE 3.4
Gender_Age_Edu categories considered as supplementary in the CA of the Gender_Age × words table. The first plane is shown. Note that the scale in this plot is quite stretched relative to that of earlier plots.

3.2.5 Implementation with Xplortext

The lines of code in `script.chap3.a.R` perform the following tasks:

- The `TextData` function builds the ALT. Responses are aggregated into six documents according to categories of the variable `Gender_Educ`. Only words used in at least 15 source documents (here: responses) are kept.

- The `LexCA` function applies CA to the ALT. The words *wife* and *husband* are declared supplementary for this analysis. No plot is asked for.
- The `ellipseLexCA` function plots the documents (= categories). Their respective positions are settled with the help of confidence ellipses.
- The `plot.LexCA` and `ellipseLexCA` functions provide us with several ways to visualize the words.
- The `TextData` function builds the same ALT as above but now this table is binded column-wise to the document categories × words table defined in terms of the 18 categories of the variable `Gen_Age_Edu`. To introduce the supplementary qualitative variable `Gen_Age_Edu` in the analysis, the argument `context.quali` has been used.
- The `LexCA` function runs the same CA as above, this time taking into account the 18 supplementary document category rows. No plots are asked for.
- The `plot.LexCA` function plots the active and supplementary categories. Added to this are the six levels of education trajectories for fixed age and gender.

3.3 Direct analysis

In this section, we present CA for the analysis of a large number of very short documents (from one to several thousand documents containing 1–20 occurrences each) which we apply to a free text answers example, analyzed without aggregation. This is therefore called *direct analysis.* The LT, with one row per non-empty free text answer, and as many columns as there are retained words, usually contains many zeros. In fact, the free text answers differ mainly in terms of the presence or absence of each of the words, and their lexical profiles are computed from a small number of occurrences—sometimes only one. The distances between such profiles are not as meaningful as they are between category profiles seen in the previous section. Nevertheless, direct analysis can be of interest, depending on the objectives and available data.

3.3.1 Data and objectives

The ***Culture_UK*** corpus gathers the answers given by the UK sample's respondents to the open-ended question: "*How would you describe the culture of your country?*" It is 9150 occurrences long and uses 1581 distinct words. The objective is to identify differences in descriptions given by the 779 (out of 1043) respondents who provided a non-empty answer. Given that the same survey was conducted in 6 other countries on similar-sized samples, another objective—not considered here—would be to compare the UK descriptions with those given in other countries.

A direct analysis of free text answers turns out to be a useful tool for characterizing types of respondents, by:

- Detecting words frequently used in the same answers, which gives information about the most-used syntagms and expressions.
- Relating the characteristics of respondents to their lexical choices, even when these relationships are complex. These characteristics can be coded as supplementary columns and projected as such onto factorial.

3.3.2 The main features of direct analysis

Each free text answer uses only a restricted number of words, leading to strong associations between small groups of answers and small groups of words. In accordance with this, eigenvalues tend to be large (in the order of 0.4–0.5) but percentages of inertia for the first axes tend to be low—around 3–5%. Typically, a slow decrease in the size of the eigenvalues is observed.

The first planes are no longer easy to interpret visually and a large number of axes now need to be examined. For each of these, high-contributing words (resp. documents) need to be identified. These may lie on only one side of the axis (positive or negative) or on both, thus corresponding to one or two groups, named *meta-keys* by Michel Kerbaol. Concerning words, within a given meta-key they are linked by syntactic and/or semantic relationships, i.e., they "work together" as a whole, indicating specific topics. In the example discussed in the following section, we find for instance the following two metakeys: (*music art theatre literature museums ballet*) and (*nothing else fish chips*). The same word may belong to several meta-keys, each time corresponding to a different meaning or nuance, due to its association with the other words, i.e., the context. Furthermore, synonyms (in the framework of the question or, more generally, within the corpus being studied) will be found in the same meta-key. Two words frequently used in the same answer will be close to each other in the factorial spaces and plots. However, two words never used in the same answer will still be close if they are frequently associated with the same "other" words. CA therefore reveals the existence of synonymous words in a non-theoretical way, simply by identifying them using similarity in terms of context.

3.3.3 Direct analysis of the culture question

In this example, words are required to be used in at least 15 documents, leading to the retention of 99 distinct words and 61.4% of the occurrences. The function words play a highly relevant role, as we will see. Individual variability, as measured by the total inertia Φ^2 (= 14.4) is large, as is also the first eigenvalue (= 0.53, see Table 3.3). However, the first eigenvalue is associated with a low percentage of inertia (3.6%). The first plane retains only 6.5% of the total inertia, which is itself spread over 98 axes. The decrease in the eigenvalues is very slow (see Figure 3.5). These are typical results.

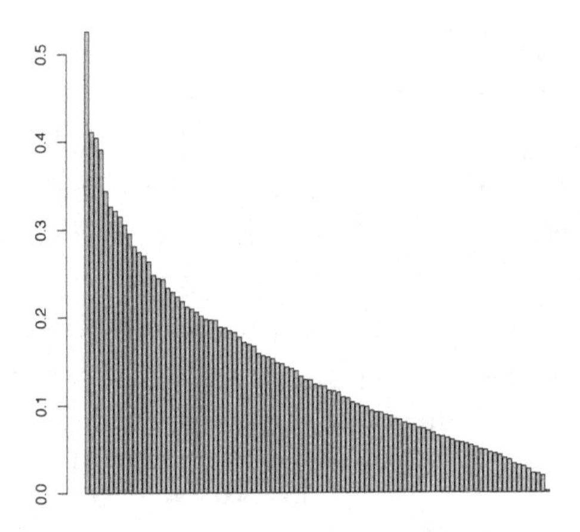

FIGURE 3.5
Bar chart of the eigenvalues.

The structure of the document and word clouds projected on the first plane has, in Figure 3.6, a quite specific shape, looking like—in the words of Jean-Paul Benzécri— a "*parabolic crescent*". This shape, commonly seen in CA, is known as the *Guttman effect*. The furthest values are projected onto the first axis on either side of the (CoG). In both clouds, the part on the left is simultaneously shorter and denser. In the cloud of documents, many points lie in the center of what would be the parabola if fitted.

When the Guttman effect is observed, the first axes exhibit an ordering of both the set of row documents (= respondents) and column words. For the respondents, this is induced by the words they use. The ordering of the words is induced by the documents that use them. The first axis ranks both documents and words according to this. In each of the two spaces, the second axis contrasts points with extreme coordinate values on the first axis with those with medium coordinate values on the same axis.

In addition, the second axis shows points which stand out from the global trend, such as for example the document points inside the curve of the parabola (such as the points "60" and "279"). According to the transition formulas, these documents use words found at the two extremes.

Reading off the words ordered by their first axis coordinate value—from negative to positive—can give meaning to the ordering (see Table 3.4). First, we read a series of function words, corresponding to a vague discourse on

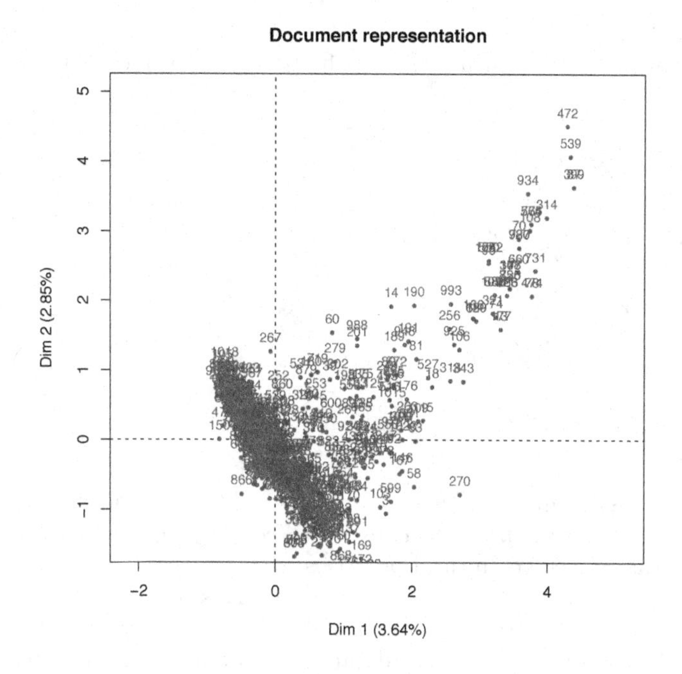

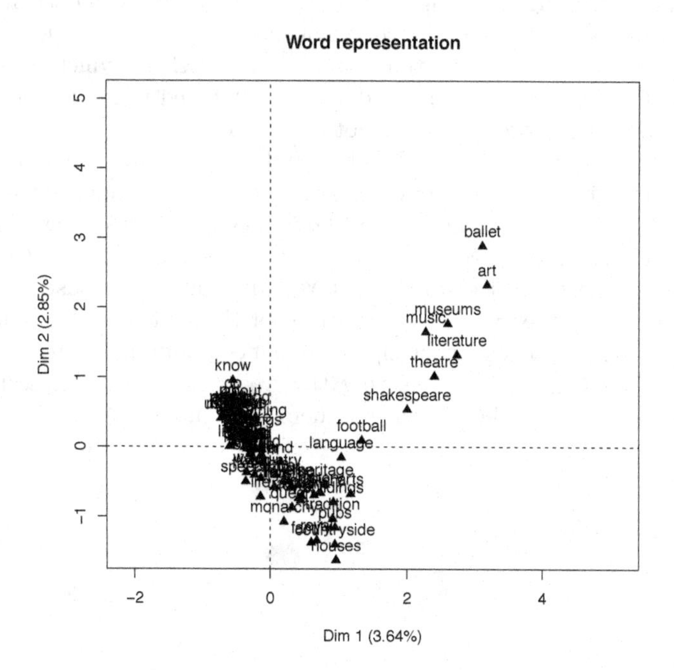

FIGURE 3.6
CA applied to free text answers to the open-ended question on culture: first factorial plane. Top: representation of the documents; bottom: representation of the words.

TABLE 3.3
Direct analysis of the culture answers: first five eigenvalues, total inertia, and Cramér's V.

```
Correspondence analysis summary

Eigenvalues
       Variance % of var. Cumulative % of var.
dim 1     0.526     3.643                3.643
dim 2     0.412     2.853                6.496
dim 3     0.404     2.802                9.298
dim 4     0.392     2.715               12.014
dim 5     0.344     2.383               14.396

Cramer's V  0.384    Inertia  14.428
```

culture such as *I can not think of anything.* Then, we find around the CoG words such as *british, freedom, chips, fish, country, monarchy, royal.* Lastly, we see words related to cultural practices.

Respondents ordered on the first axis in terms of the words they use are therefore ordered according to their ability to describe the culture of their country. A strongly negative coordinate corresponds to a low capacity of response. Then, moving to the right, their capacity increases, and the main parts of the culture are mentioned by the most extreme positive-valued responses. This has been deduced from the responses themselves, which is a particularly remarkable result. As for the documents located inside the curve of the parabola, they use words found at both extremes.

In order to better-interpret the plots and improve our understanding, one approach intuitively comes to mind: cluster together respondents who are close to each other on the plane, i.e., those who use a similar vocabulary. Then, each cluster will be characterized by the free text answers closest to its CoG. There are many different ways to form clusters. One way is to construct them in terms of a set of intervals in the direction of the first axis. In Figure 3.7, a division into eight homogeneous clusters has been obtained using a clustering algorithm (see Chapter 4). Selecting the free text answers closest to each cluster's CoG can then help interpret them (see Table 3.5).

TABLE 3.4
Semantic words whose coordinate value on the first axis is above 0.5 or below -0.5. Coordinate values, contributions, and the representation quality of each word, are shown.

	coord	contrib	cos2
art	3.18	11.62	0.21
ballet	3.11	5.23	0.11
literature	2.73	8.61	0.18
museums	2.60	5.72	0.11
theatre	2.40	11.32	0.26
music	2.27	13.13	0.25
shakespeare	2.00	3.52	0.07
education	1.65	3.15	0.04
football	1.34	1.21	0.03
arts	1.17	2.85	0.08
language	1.04	0.58	0.02
houses	0.96	0.57	0.01
pubs	0.95	0.46	0.01
countryside	0.95	1.43	0.03
buildings	0.92	0.95	0.03
tradition	0.91	0.73	0.02
heritage	0.81	0.99	0.03
history	0.74	1.12	0.04
royal	0.68	1.01	0.04
english	0.64	0.35	0.01
family	0.60	1.00	0.03
-----			
i	-0.51	1.48	0.09
has	-0.51	0.14	0.01
best	-0.51	0.18	0.01
it	-0.53	1.21	0.06
too	-0.53	0.20	0.01
to	-0.53	1.31	0.04
am	-0.54	0.24	0.01
do	-0.54	0.81	0.04
know	-0.55	0.32	0.02
we	-0.56	1.12	0.06
be	-0.56	0.50	0.02
very	-0.58	0.51	0.02
not	-0.59	2.25	0.12
live	-0.59	0.27	0.01
you	-0.61	0.41	0.02
there	-0.62	0.22	0.02
lot	-0.62	0.34	0.02
what	-0.64	0.41	0.03
much	-0.66	0.54	0.02
so	-0.66	0.28	0.02
one	-0.68	0.27	0.02
used	-0.72	0.32	0.02

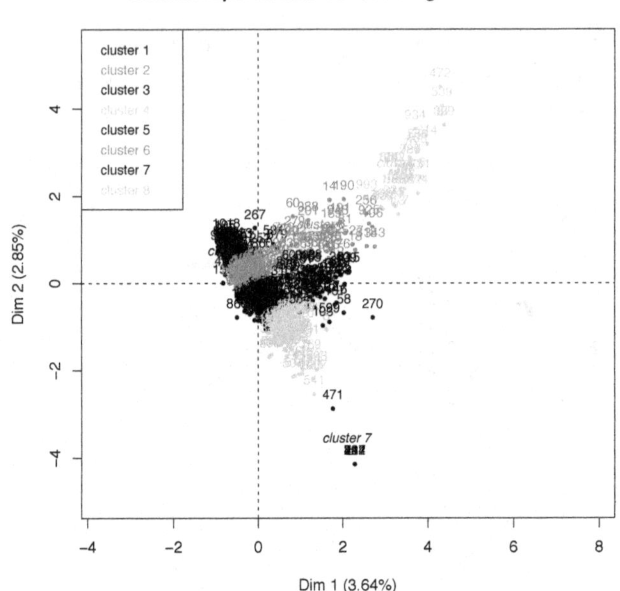

FIGURE 3.7
Division of the respondents cloud into eight clusters in terms of ability to describe their country's culture.

3.3.4 Implementation with Xplortext

The lines of code in `script.chap3.b.R` perform the following tasks:

- The `TextData` function builds the LT using column 11 (culture question). Only words found in at least 15 documents (source documents, which are also the documents analyzed here) are retained.
- The `summary.TextData` function provides some initial statistics on the corpus.
- The `LexCA` function applies CA to the LT. No plot is requested.
- The `summary.LexCA` function provides a summary of the CA results.
- The `plot.LexCA` function is used to visualize the eigenvalues.
- The `plot.LexCA` plots the documents and words on separate graphs.
- Various R commands are used to obtain and output the list of words ordered by their first axis coordinate values.
- The function `LexCA` is run again but only keeping two axes.

- The `LexHCca` function groups the documents into eight clusters. Each of these is then described in terms of the documents closest to its CoG, called *paragons* in the **Xplortext** package.

TABLE 3.5
Description of the 8 clusters in terms of the three free text answers closest to their CoGs.

Cluster	Size	Free text answers closest to the CoG
`Cluster 1`	164	– I do not think very much of British culture at all. – it is getting more violent, muggings etc., like America. – can not think of anything, my mind is a blank at this stage.
`Cluster 2`	192	– really, I always think of someone British is reserved and fair, old stiff upper lip, quiet honest and fair, the match, the captain would not cheat, second fairly than win cheating – we have not got much left except history and royalty – I am thinking of the arts, I suppose the British way of living, slightly upper class, not plebeian, that is all
`Cluster 3`	150	– a punk, fish and chips, rain, nothing else – beer, steak and kidney pies, fish and chips. – royalty, legal system, police force is envied throughout the world, nothing else
`Cluster 4`	103	– Sunday roast, royal family, the arts, nothing else. – countryside, green fields and trees, wildlife. – the royal family, cotton mills, the northern heritage.
`Cluster 5`	50	– literature, history and old buildings, art, higher education for older people. – theatre, Shakespeare, museums, red buses and black taxis, beefeaters, Buckingham palace (other famous buildings and places), the royal family. – painting, arts and crafts, opera, Shakespeare.
`Cluster 6`	36	– gardens, opera, ballet and the theatre and pottery – music, middle of the road music, reading, sport, especially football – art, building, heritage in general.
`Cluster 7`	8	– education. – education. – leisure activities, personal health (health clubs), work conscious, education conscious, financially aware
`Cluster 8`	38	– museums, art galleries, William Shakespeare, poetry. – literature, music, rotten design – literature, music, athletics.

4

Clustering in textual data science

4.1 Clustering documents

The point of clustering a set of documents is to divide them into groups called clusters—not chosen *a priori*—so that the documents' lexical profiles are similar within the same cluster, and differ significantly from one cluster to another. This clustering must take into account all of the retained vocabulary, i.e., it requires a multidimensional approach to cluster construction. One of four main approaches is typically used:

- *Direct partitioning*: from what is known about the documents (which may come from an earlier CA and the study objectives, we pre-select the number Q of clusters to build, then select Q documents which cover the diversity found in the whole set. Then, the other documents are grouped around these, based on similarity. The initial documents may be randomly selected to begin with, and iteratively recalculated thereafter.

- *Divisive hierarchical clustering*: we divide the whole set of documents into two clusters, so that each is both as homogeneous and different from the other as possible in terms of vocabulary. Then, each is again divided in two, according to the same principle. In general, the process stops when the obtained clusters are sufficiently homogeneous with respect to the chosen criterion. This method leads to a succession of nested partitions, usually represented in the form of a *hierarchical tree*, also known as a *dendrogram*.

- *Agglomerative hierarchical clustering (AHC)* : the two documents closest to each other from a lexical point of view are grouped together to form a new document. Iteratively, we then search for the two closest documents and group them, until we have only one document left. Here again, we obtain a succession of nested partitions which can be represented as a hierarchical tree (see Figure 4.1).

- *Agglomerative hierarchical clustering with contiguity constraints*: the procedure is similar to that followed in agglomerative hierarchical clustering but, here, only contiguous documents can be aggregated. In our context, contiguity is usually in terms of chronology. The obtained succession of nested partitions can again be represented as a hierarchical tree which takes into account the underlying hierarchical structure of the corpus (see Figure 4.4).

Clustering in its various forms is exploratory in nature and comes without prior notions of what to expect. In this book, we are not focused on class prediction (categorizing documents), also known as *classification* and *discrimination*, which consists in attributing documents to classes they are thought to belong to. This could be for instance when we want to predict a document's author or date, both interesting goals but out of the scope of the exploratory analyses presented here.

Clustering methods can be used for very different purposes, such as automatically grouping works (or authors) in order to contrast them with a grouping made by experts, defining lexically homogeneous time intervals in a sequence of documents issued over time, uncovering the hierarchical structure of parts of a corpus or a single text, organization of a textual database by theme in order to accelerate searches corresponding to queries, and clustering free text answers, as already outlined in Chapter 3.

4.2 Dissimilarity measures between documents

Whatever the clustering method, a measure of dissimilarity between statistical units (here, documents) needs to be defined. This dissimilarity may or may not be a distance measure in the mathematical sense of the term. In addition, any hierarchical clustering requires an aggregation method. These choices are important because they affect the results. Moreover, the joint use of clustering and factorial methods forces the choice of dissimilarity measures between documents used by each to be coherent with each other. Recall that CA projects documents initially located in a vector space endowed with a χ^2 distance into a vector space endowed with the classical Euclidean distance. If all axes are retained, the Euclidean distance between two documents calculated using their factorial coordinates is equal to the χ^2 distance in the initial space.

Although clustering methods are flexible and can work with a range of distances or dissimilarity measures, in this chapter we only consider the case where the document points are placed in a Euclidean space coming from an earlier factorial analysis. The I documents are thus referenced by their coordinates on the S retained factorial axes, noted $F_s(i), \{s = 1, \ldots, S\}$, $\{i = 1, \ldots, I\}$, which leads to the use of the classical Euclidean distance between documents.

4.3 Measuring partition quality

4.3.1 Document clusters in the factorial space

After clustering, the document cloud N_I is partitioned into Q clusters labeled by q, $\{q = 1, \ldots, Q\}$, with respectively I_q members each. Documents in cluster q are labeled by i_q, $\{i_q = 1, \ldots, I_q;\ q = 1, \ldots, Q\}$. The factor on the row documents, $F_s(i_q)$, $\{s = 1, \ldots, S\}$, places them on the CA axes. They are attributed the weights f_{i_q} also coming from CA or, in some cases, multiple factor analysis for contingency tables (MFACT), described in Chapter 6.

In the factorial space, each of the clusters is represented by its center of gravity (CoG) C_q $\{q = 1, \ldots, Q\}$, and attributed the weight f_q $\{q = 1, \ldots, Q\}$ equal to the sum of the weights of the documents in it. The C_q factorial coordinates $F_s(C_q)$, $\{s = 1, \ldots, S\}$ are the weighted means of the coordinates $F_s(i_q)$, $\{i_q = 1, \ldots, I_q\}$ of the corresponding cluster's documents. The document cloud and the set of cluster CoGs are centered on the global CoG.

4.3.2 Partition quality

We want to partition the set of documents so that within each cluster, the documents use similar vocabulary, and from one cluster to another, the vocabulary is relatively different. The clustering algorithm has to work with the set of retained words, often numerous, and aggregate documents in a way that achieves both intra-cluster homogeneity and good separation between clusters. This goal suggests that we evaluate partition quality by means of a criterion which takes both into account, using within-cluster and between-cluster inertia.

The total inertia of the document cloud N_I can be broken down into within-cluster and between-cluster inertia according to Huygens theorem of inertia decomposition. As the axes of the factorial space are orthogonal, the inertias can be calculated as sums of the inertias of each axis. As both the document and CoG clouds are centered, the inertia can be decomposed as:

$$\sum_{q=1}^{Q}\sum_{i=1}^{I_q}\sum_{s=1}^{S} f_{i_q} F_s^2(i_q) = \sum_{q=1}^{Q}\sum_{s=1}^{S} f_q F_s^2(C_q) + \sum_{q=1}^{Q}\sum_{i=1}^{I_q}\sum_{s=1}^{S} f_{i_q}(F_s(i_q) - F_s(C_q))^2$$

Total inertia = between-cluster inertia + within-cluster inertia.

For a fixed number of clusters Q, the smaller the within-cluster inertia, the better the partition quality. This is because a small within-cluster inertia means that documents with similar lexical profiles have been placed in the same clusters. As the total inertia is constant, the smaller the within-cluster inertia, the larger the between-cluster inertia becomes. Thus, for fixed Q, minimizing the within-cluster inertia is equivalent to maximizing the between-cluster inertia. This leads, for a fixed Q, to the following ratio as a measure of partition quality:

$$\frac{\textit{between-cluster inertia}}{\textit{total inertia}}.$$

This takes values between 0 (no between-cluster inertia, i.e., all cluster CoGs are the same and equal to the overall CoG) and 1 (the between-cluster inertia is equal to the total inertia, i.e., all of the documents in each cluster have the same factorial coordinates and thus the same lexical profiles). This ratio corresponds to the percentage of the total variability represented by the given partition. For a fixed Q, it turns out that optimizing this criterion has a prohibitory calculation cost, meaning that it is unachievable in practice. However, it does guide the strategy we describe in what follows.

4.4 Dissimilarity measures between document clusters

Constructing a hierarchy or dendrogram requires a between-cluster dissimilarity measure, also known as a *aggregation method*. In any case, no matter the method used for building clusters, it is always useful to be able to measure the dissimilarity between them.

4.4.1 The single-linkage method

In the *single-linkage* method, the dissimilarity between clusters q and q' is equal to the smallest distance between a document in q and a document in q'. This criterion can lead to low density and spread-out clusters as it is susceptible to a "chaining" effect.

4.4.2 The complete-linkage method

In the *complete-linkage* method, also known as the *diameter*, the dissimilarity between clusters q and q' is equal to the largest distance between a document in q and a document in q'. This leads to concentrated (low-diameter) clusters.

4.4.3 Ward's method

When clustering documents represented by weighted points in a Euclidean space—as is the case here, other between-cluster dissimilarity measures can be used. We can for instance compute the coordinates of the cluster CoGs on the factorial axes and measure the dissimilarity between clusters in terms of the *distance between their CoGs*.

We focus here on clustering associated with a factorial method such as

CA. In this case, we prefer to take the inertia into account. The dissimilarity between two clusters is measured by the decrease in between-cluster inertia that their aggregation produces. This criterion (or method), called *Ward's method* or the *minimum variance method*, is denoted δ and computed as we now describe.

Suppose that clusters q and q', with CoGs C_q and $C_{q'}$, are attributed weights f_q and $f_{q'}$, corresponding to the respective sums of the weights of the points belonging to each cluster. Let us denote the squared Euclidean distance between their CoGs by $d^2(C_q, C_{q'})$. If the two clusters are aggregated, the between-cluster inertia decreases by:

$$\delta(q, q') = \frac{f_q \cdot f_{q'}}{f_q + f_{q'}} d^2(C_q, C_{q'}).$$

Ward's method is often used in unconstrained hierarchical clustering since it tends to lead to homogeneous and well-separated clusters. Nevertheless, the local optimization used here to create each new node does not necessarily lead to the globally optimal partition for a given, fixed number, of clusters.

Remark:

These types of node aggregation can be used even when one or both nodes are singletons. All of the methods presented here are equivalent to the Euclidean distance when applied to two singletons.

4.5 Agglomerative hierarchical clustering

Agglomerative hierarchical clustering (AHC) starts with the elements to cluster—here documents—as terminal nodes. The underlying algorithm for building a hierarchical tree is the same, apart for the actual aggregation method chosen.

4.5.1 Hierarchical tree construction algorithm

At initialization, we have as many clusters as there are documents. Each of them, also known as terminal node, is made up of one document and is called a *singleton*. Then:

1. Construct the dissimilarity matrix between the I terminal nodes. Initially, the entry corresponding to a given row and column is equal to the Euclidean distance between the corresponding terminal nodes. It is thus sufficient to fill in only the upper or lower diagonal of the matrix; by convention, we fill in the lower.

2. The two nodes closest to each other, whether terminal or not, are aggregated into a new node, numbered between $I + 1$ (first non-terminal node) and $2I - 1$ (last non-terminal node). We index each node created by the algorithm with the value of the aggregation criterion between its two sub nodes. The resulting tree is known as a *hierarchical index tree* or *indexed hierarchy*.
3. After each fusion of two nodes, the dissimilarity matrix needs to be updated. The rows and columns corresponding to the aggregated documents are removed, and a new row and column are added, corresponding to the newly formed node. Before iterating, the entries corresponding to this row and column are updated by calculating the dissimilarities between this new node and all other remaining nodes.
4. While we still have more than one cluster, return to Step 2. When all of the documents have been aggregated together as the root node, i.e., the ($2I$-1)th node, the tree is complete.

By construction, this is a binary tree, which means that each non-terminal node brings together two sub nodes. Nodes formed are therefore nested, right up to the largest one, which contains all documents. To get a better understanding of this algorithm, see the example in Section 4.9, including Figure 4.1 and Table 4.2.

In practice, however, faster algorithms are nearly always used, such as the nearest-neighbor chain algorithm, or the algorithm used in the **flashClust** R package.

4.5.2 Selecting the final partition

We can *cut* a hierarchical tree by drawing a horizontal line across it. Each such line defines a partition (see Figure 4.2), and drawing it at different heights corresponds to different partitions. In fact, a given partition will be nested in every partition made further up the tree. A hierarchical tree thus corresponds to a sequence of nested partitions, from the most extreme (each terminal element corresponds to a one-document cluster) to the least (there is one cluster containing all documents).

4.5.3 Interpreting clusters

An interpretation of clusters in terms of under- or over-used words can help to understand the semantic content of each of them. This can then be built upon by studying the relationship between the partition (and the clusters) and contextual variables, whether quantitative or qualitative. This can lead to a better understanding of significant connections between contextual characteristics of documents and the range of vocabulary found in each cluster. Methods for interpreting clusters can be found in Section 4.11.

4.6 Direct partitioning

We can also directly partition the documents by clustering them around centers. This type of partition construction is cheap in calculation time, meaning it can be used on extremely large datasets, unlike hierarchical clustering whose calculations become prohibitive when there are a large number of elements to cluster (tens of thousands or more). Many variants exist. Here we limit ourselves to presenting the well-known k-means algorithm.

Just like hierarchical clustering, we suppose that documents are represented by their factorial coordinates coming from an earlier analysis. We then make a decision that S axes (perhaps all of them) are to be retained for subsequent analyses. The distance between documents is then given by the Euclidean distance. It is necessary to set the number of clusters Q before starting. The ratio of the between-cluster to total inertia is then adopted as a measure of the quality of a given partition into Q clusters.

At the initialization step, a first partition P^0 is obtained by selecting Q initial CoGs. These CoGs usually correspond to either Q documents randomly selected or due to them being considered good representatives of the set's variability. Then, each document is assigned to the cluster whose CoG it is closest to. The value of the ratio of inertia, noted η^0 at the initial step, decreases at each iteration:

$$\eta^0 = \frac{\textit{between-cluster inertia}}{\textit{total inertia}}.$$

Then, at the kth iteration of the algorithm:

1. Calculate the CoG of each cluster in the partition P^{k-1}. These CoG are then taken as the new cluster centers.
2. Then, again, associate each document to the closest center, leading to the partition P^k.
3. Calculate the ratio of inertia η^k.
4. If $(\eta^k - \eta^{k-1}) < \textit{threshold}$, output the partition P^k. Otherwise, iterate from Step 1 until this is true or a pre-defined maximum number of iterations has been performed. Note that both the threshold and the maximum number of iterations have default values in the packages being used; these can be modified by users.

It can be shown that after each iteration of this algorithm, the value of the ratio of inertia decreases, meaning that it converges, usually very quickly. Even if the maximum number of steps is fixed at 10, which is a typical value, the algorithm has often already converged before this number of iterations has been reached.

4.7 Combining clustering methods

Direct clustering methods, in comparison to AHC, have two interesting features:

- optimization of a criterion that is, in a sense, defined on the whole data set at once. However, the best possible partition into Q clusters in terms of the value of the ratio of inertia will not necessarily be found since the minimum obtained is still local, rather than global.
- reduced calculation time.

On the other hand, they require the number of clusters to be chosen before starting, whereas AHC, with its hierarchical tree, provides useful information on this point. This suggests using a mixed strategy, bringing together hierarchical clustering and direct partitioning to combine, as much as possible, each of their strengths.

4.7.1 Consolidating partitions

Generally speaking, the partition given by an AHC for a given number of clusters does not optimize the ratio of inertia criterion either. What we can do is apply to this partition the direct partitioning algorithm presented above, which will improve the clustering in terms of the criterion and given more homogeneous and better-separated clusters. Of course, in doing so, we lose the connection between the final partition and the original hierarchy.

4.7.2 Direct partitioning followed by AHC

When clustering a very large number of documents, one strategy is to first do direct partitioning with a large number of clusters (from 100 to 1000 depending on the total number of documents). This acts as an intermediate step, involving too many clusters to be useful or interpretable. Despite this, the clusters are homogeneous. With this in mind, we can then run an AHC taking these clusters as terminal nodes, with weights equal to the sum of the documents' weights in each of them. The indexed hierarchy constructed in this way should not be very different to the one we would have had if starting with all of the documents individually.

4.8 A procedure for combining CA and clustering

The joint use of CA and AHC can lead to mutually enriched results. The starting point is a matrix of (row) documents × (column) words. The document cloud on the one hand, and the word cloud on the other, sit in spaces endowed with χ^2 distances. The following steps are then performed:

1. The CA transports these two clouds to spaces endowed with the Euclidean distance, thus revealing similarities between documents, and similarities between words. Furthermore, the set of distances between words takes into account the vocabulary structure in the corpus. Superimposing the two representations helps to visualize links between words and documents.
2. A clustering of the documents in this Euclidean space can be performed using their coordinate values on the factorial axes considered interpretable.
3. Document clusters can then be described using all available information, including under- and over-represented words in each cluster, characteristic documents, and all of the contextual variables used as supplementary variables in the CA.
4. Next, the CoGs are placed on the factorial axes.
5. Then, the hierarchical tree is also represented on the plane.

In this way, contrasts that appear on the factorial axes can be related to the hierarchical structure of the corpus, and the partition obtained is more meaningful when the clusters are shown in the factorial planes.

4.9 Example: joint use of CA and AHC

4.9.1 Data and objectives

Starting from the *Life_UK* corpus, already used in earlier chapters, 6 documents are formed corresponding to aggregated responses from the following age groups: 18_24, 25_34, 35_44, 45_54, 55_64, and ≥65. These 6 age groups correspond to the 12 age groups in the database, grouped into time-adjacent pairs. The 135 words used in at least 15 documents are retained. In terms of the number of documents being considered, this is essentially a small toy example. Here, the documents × words table is an ALT.

4.9.1.1 Data preprocessing using CA

First, we run CA on the age groups × words table (see Table 4.1). The document categories are then given in terms of their coordinate values on the five factorial axes (the full space) and their CA weights. Figure 4.1 shows the distances between these documents on the first factorial plane. In the full space, the total inertia of the document cloud is equal to the sum of the eigenvalues, which is 0.1154. The first decision to make is how many axes to retain before clustering. Here, we keep all five. In this toy example, the results are in any case similar when retaining 3, 4, or 5 axes.

TABLE 4.1
CA of the age groups × words table. Initial results are shown.

```
Correspondence analysis summary

Eigenvalues
       Variance % of var. Cumulative % of var.
dim 1    0.0445   38.5416              38.5416
dim 2    0.0328   28.4456              66.9873
dim 3    0.0148   12.8400              79.8272
dim 4    0.0129   11.1496              90.9769
dim 5    0.0104    9.0231             100.0000

Cramer's V  0.1519     Inertia  0.1154
```

4.9.1.2 Constructing the hierarchical tree

The document categories are the terminal nodes of the tree, numbered 1 (18_24) to 6 (≥65). This is the most extreme partition, with every cluster made up of a single document category. At this level of the tree, the within-cluster inertia is zero and the between-cluster inertia equals the total inertia—here 0.1154.

After each hierarchical aggregation of two nodes, the between-cluster inertia decreases (and the within-cluster inertia increases) by an amount equal to the dissimilarity between the two nodes joined (in terms of Ward's method).

At the top of the hierarchy, the between-cluster inertia is zero and the within-cluster inertia is equal to the total inertia (which remains the same throughout the process). The value attached to each node is equal to the value of the dissimilarity between the two clusters that were aggregated in forming it.

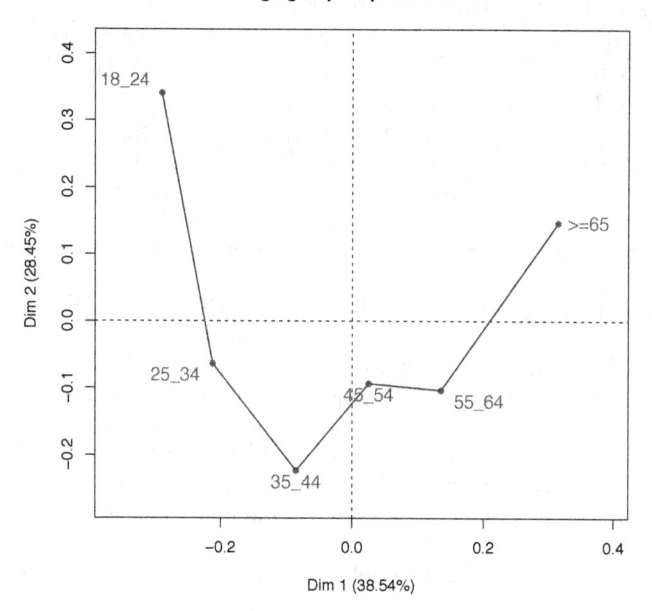

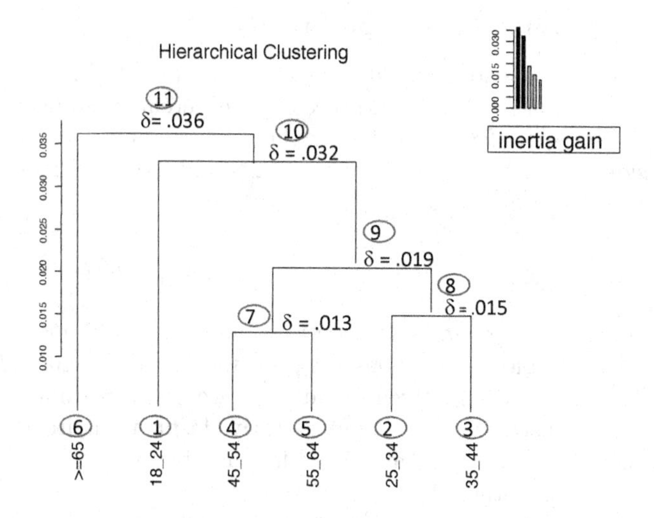

FIGURE 4.1
CA and clustering of the six age groups. Top: categories mapped on the first CA plane; bottom: indexed hierarchy. The numbers (circled) identifying nodes, and the successive values of Ward's method have been manually added. Above the tree, a small diagram shows the progression in the value of the aggregation criterion, here equal to gains in between-cluster inertia.

TABLE 4.2
Values associated with the hierarchy shown in Figure 4.1.

Node	q	q'	δ	in %	Within-cluster	Between-cluster
7	45–54	55–64	0.0128	11.1	0.0128	0.1026
8	25–34	35–44	0.0150	13.0	0.0278	0.0876
9	7	8	0.0190	16.5	0.0468	0.0686
10	18–24	9	0.0324	28.1	0.0792	0.0362
11	over 64	10	0.0362	31.4	0.1154	0.0000
Sum			0.1154	100.00		

These results are summarized in Table 4.2. Note that:

- the sum of the indices δ is equal to the total inertia of the document cloud in the space corresponding to the retained axes. This is always true.
- the value of the index δ increases (or at least does not decrease) as the tree is built. Therefore, it is never the case that the index associated with a node is less than the one associated with a node created before it.

In this example, the indexed hierarchy in Figure 4.1 seems to be well balanced.

4.9.1.3 Choosing the final partition

The split into two clusters would express 31.4% of the total inertia. Then, the `18-24` group breaks off, also with a clearly different vocabulary to the rest. This split represents 28.1% of the inertia, almost as much as the first. Thus, the division into three clusters ($\geq$`65`, `18-24`, and `25-64`) corresponds to well over half of the total inertia:

$$\frac{\textit{between-cluster inertia}}{\textit{total inertia}} = \frac{0.0686}{0.1154} = 0.595.$$

The corresponding *cut*, located between nodes 9 and 10, means not making the last two possible aggregations (i.e., not forming nodes 10 and 11).

It is often more informative to read a hierarchical tree from top to bottom, even if it was constructed in an agglomerative fashion. Figure 4.1 supports us in asserting that there is a marked difference between vocabulary used by those $\geq$`65` and the others.

Notice that the ratio of inertia of the partition so far is equal to the sum of the inertias at each split: 59.5%=31.4%+28.1%. This split into three clusters is the one automatically output by the **LexHCca** function, which in part depends on the minimum and maximum number of clusters allowed by default. It turns out that here, both are equal to three. The former is the default value, while the latter is the minimum of 10 and the number of documents divided by two ($6/2 = 3$ here).

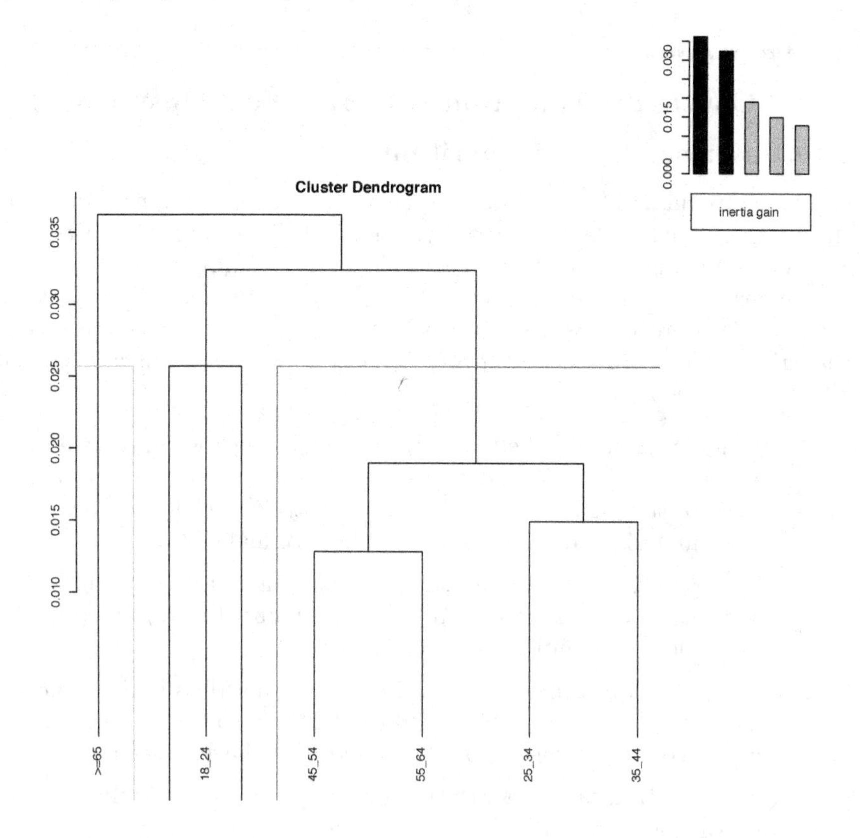

FIGURE 4.2
The tree cut chosen for the tree in Figure 4.1. A partition into 3 clusters is obtained.

Continuing the interpretation, moving further down the tree we see that subsequent nodes aggregate age categories which are close to each other (according to the indices associated with the nodes). Here, it is clear that 3 clusters is indeed the best choice because the vocabulary of the `18_24 years` group is almost as different to that of the `25-64 years` group as the latter is to the $\geq$65 group.

Remark:

A division into clusters of contiguous age groups occurs here without actually imposing this constraint. We will see in Section 4.10.2 that for the same data, using shorter age ranges does not in fact lead to chronological ordering if a contiguity constraint is not added.

4.10 Contiguity-constrained hierarchical clustering

4.10.1 Principles and algorithm

It may be useful and/or necessary to include—when performing hierarchical clustering of documents—a contiguity constraint. This is usually in terms of chronology, and means that nodes in the hierarchy can only aggregate together contiguous documents or clusters. To include this constraint in a clustering algorithm, we proceed as described below. As when performing hierarchical clustering without constraints, at Step 0, we start with every document as its own cluster. Then:

1. A similarity matrix between the terminal nodes is constructed in terms of the Euclidean distance between documents. The (i, i')th entry of this matrix is equal to $d_{max} - d(i, i')$, where d_{max} is the maximum distance between two of the terminal nodes.
2. A contiguity matrix is constructed. This symmetric matrix has the elements being clustered as rows and columns. Its (i, j)th entry is 1 if i and j are contiguous, and 0 otherwise.
3. The Hadamard product between these two matrices is then calculated, i.e., the element-wise product. In the resulting matrix, only entries corresponding to connected nodes will have non-zero values.
4. In this matrix, we look for the largest value; the corresponding nodes are then aggregated to form a new one.
5. The similarity matrix, in the complete-linkage sense, is then updated. Rows and columns corresponding to newly grouped documents are removed, and a new row and column are added corresponding to the new node from Step 4. The similarity values between this new node and all remaining nodes are then calculated.
6. The contiguity matrix is updated. Rows and columns corresponding to newly grouped documents are removed and a new row and column are created corresponding to the new node. Contiguity between this node and the remaining nodes is again signaled by a 1, and non-contiguity by a 0.
7. While all documents are not found in the same node, return to Step 4, else: stop.

The tree is by construction binary, and the clusters are nested, right up to the $(2I - 1)$th node, containing all of the documents.

4.10.2 AHC of age groups with a chronological constraint

Let us return to the *Life_UK* example, now starting with the 12 original age groups and the objective of using AHC with a chronological contiguity constraint (AHCCC) to perform clustering. This is done starting with the factorial coordinates of the age groups calculated from an earlier CA. A quick look at the axes shows that only the first two are interpretable in the sense of similarities and differences between age groups (see Figure 4.3).

This AHCCC groups together more homogenous age groups—in terms of vocabulary—than those obtained in Section 4.9, where *a priori* aggregation of contiguous age groups was performed, so structure in the vocabulary is better taken into account here. However, clusters now have a larger range of sizes.

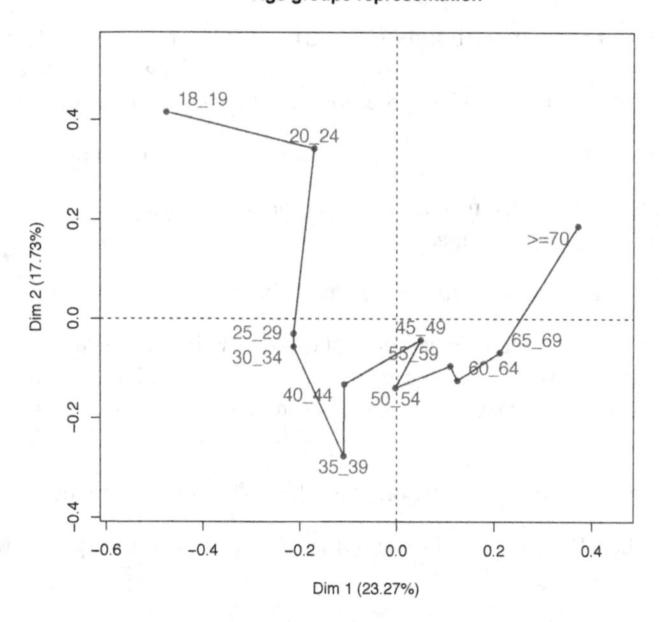

FIGURE 4.3
CA of age groups × words table. The age groups are shown in the (1,2) plane.

As we do not want too few clusters, we request at least four. The automatically suggested cut actually leads to six clusters (see Figure 4.4—top). In accordance with the constraint, only contiguous nodes have been aggregated.

This tree cut includes three clusters which have only one element each, corresponding to the two youngest and the oldest age groups (see Figure 4.4). This is coherent with what we see in the first factorial plane, as expected, since the first two axes were retained for the clustering. This suggests that the clusters found represent clearly distinct vocabulary. Note that this partition into six clusters could then be used as a new categorical variable.

This tree can be compared with the one built using AHC, i.e., without constraints (see Figure 4.4—bottom). In the latter, we see that the `45-49` age group is aggregated with non-contiguous ones, such as the `55-70` node under a 6-cluster partition. As this cluster is unlikely to correspond to interesting structural information, it is not of much interest to us.

4.10.3 Implementation with Xplortext

The script named `script.chap4.a.R` performs the following tasks:

- A new variable `Age_en_Classes_B`, reducing the number of age group clusters to six, is formed.
- The `TextData` function builds the ALT. Free text answers are aggregated into six documents according to the categories of `Age_en_Classes_B`. Only words found in at least 15 source documents (= responses) are retained.
- The `LexCA` function runs CA on the ALT. No plots are requested.
- The `plotLexCA` function plots the documents (= age groups), as well as the trajectories in terms of age.
- The `LexHCca` function runs an unconstrained AHC.
- The `TextData` function constructs the ALT which aggregates free text answers according to the 12 categories of the `Age_en_Classes` variable. Only words found in at least 15 source documents (= free text answers) are retained.
- The `LexCA` function runs CA on the ALT. No plots are requested.
- The `plotLexCA` function plots the documents (= age groups) as well as the trajectories in terms of age.
- The `LexCHCca` function runs an AHC with contiguity constraints.
- The `LexHCca` function runs an unconstrained AHC.

4.11 Example: clustering free text answers

4.11.1 Data and objectives

We now return to the *Life_UK* corpus to cluster the 1043 individuals of the UK sample starting from their free text answers.

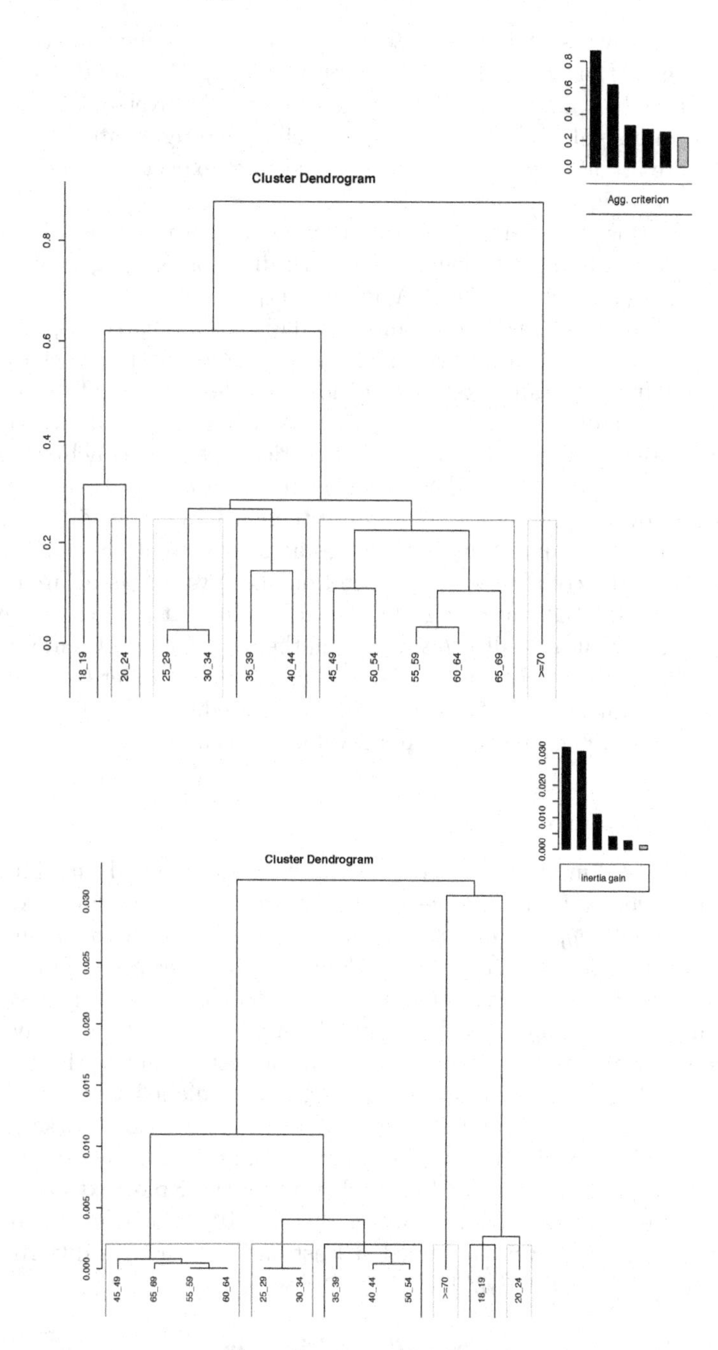

FIGURE 4.4
Clustering the 12 age groups. Top: tree built using AHCCC; bottom: tree built using AHC.

The objective is to build clusters of individuals similar to each other in terms of lexical profiles. In what follows, the words *documents*, *free text answers* and *individuals* are used interchangeably. The diverse themes that can be found in a single free text answer, as well as the tight interconnection between content and form, mean that we should not expect clusters with simple meanings or unique themes.

Despite this, the partitioning of respondents into groups will be useful because each will be at the same time markedly more homogeneous than the whole set, and of smaller size. A joint interpretation of the results of the clustering and the CA is of great interest. This AHC greatly helps to decrypt the lexical content of each cluster, thanks to the extraction of characteristic words and free text answers. In addition, the description of clusters using contextual variables can show connections between them and lexical choices.

For these tasks, we will perform statistical tests to conclude as to the significance either of the under- and over-represented words in clusters, the under- and over-represented categories in the clusters, or the difference in the mean of contextual quantitative variables across different clusters.

As usual, the conclusion will depend on the p-values associated with the observed value of the statistic. Note that the **Xplortext** package complements p-values with what are called *test values* in the terminology of Ludovic Lebart. A test value can be seen as a standard normal variable under the null hypothesis corresponding to the statistical test that is used. A test value either below -1.96 or over 1.96 corresponds to a p-value less than 0.05.

4.11.2 Data preprocessing

Only words used in at least 15 free text answers are retained. In addition, tool words are deleted. To do this, we use the list provided by the **tm** package. Also, the words *anything, nothing, else, can, think*, used in the expressions *nothing else*, *anything else*, and *I can not think of anything else*, are removed because they push free text answers that use them too close together, even though the respective meaning of each is quite different. The words *can* and *not* are written separately in the corpus in order to correctly calculate the occurrences of each one. Lastly, the word corresponding to the unique letter *s*, which comes from the possessive *'s* isolated after removing punctuation marks (and thus the apostrophe), is eliminated. Note that the English language stoplist can be output by typing `stopwords("en")` when using the **Xplortext** package.

The analyzed documents $\times$ words LT has 1039 rows, corresponding to individuals whose answer contained at least one of the retained words, and 78 word columns (see Table 4.3).

4.11.2.1 CA: eigenvalues and total inertia

First, we run CA on this relatively sparse LT. This brings up some peculiarities compared with CA on an ALT. The eigenvalues, and thus the inertia, are

TABLE 4.3
Extract of results given by the `summary.TextData` function.

```
TextData summary
                        Before   After
Documents              1043.00 1039.00
Occurrences           13917.00 5026.00
Words                  1334.00   78.00
Mean-length              13.34    4.84
NonEmpty.Docs          1040.00 1039.00
NonEmpty.Mean-length     13.38    4.84

Index of the  20  most frequent words
        Word Frequency N.Documents
1  family          705         624
2  health          612         555
3  good            303         254
4  happiness       229         218
5  money           172         169
6  life            161         149
7  job             143         137
8  happy           137         123
9  children        131         129
10 work            118         113
11 friends         116         111
12 husband          96          92
13 home             90          89
14 live             84          76
15 peace            79          75
16 wife             76          72
17 living           68          66
18 enough           68          61
19 people           63          55
20 able             56          51

Summary of the contextual categorical variables
   Gender        Education        Gender_Educ    Gender_Age
 Man  :494   E_Low   :478   M_EduLow :221   M<=30 :134
 Woman:545   E_Medium:418   M_EduMed :197   M31_55:193
             E_High  :143   M_EduHigh: 76   M>55  :167
                            W_EduLow :257   W<=30 :132
                            W_EduMed :221   W31_55:251
                            W_EduHigh: 67   W>55  :162

Summary of the contextual quantitative variables
      Age
 Min.   :18.00
 1st Qu.:30.00
 Median :44.00
 Mean   :45.82
 3rd Qu.:61.00
 Max.   :90.00
```

elevated, as is Cramér's V (see Table 4.4). The decay in the sequence of eigenvalues is very slow and the percentages of total inertia associated with the first axes are very low. This means that the factorial axes could be unstable in the face of small changes in the data. The first planes do not therefore offer a good visual summary like in the case of the CA of aggregate tables, but rather, successively uncover various word associations (= meta-keys) used by groups of respondents.

As a consequence, instead of studying the axes in isolation, it will be beneficial to study the subspace corresponding to the axes considered to be informative, using clustering. The description of clusters of the retained partition will invoke both the lexical content of the clusters and the main characteristics of the respondents who are in them.

TABLE 4.4
Eigenvalues, inertia, and Cramér's V.

```
Correspondence analysis summary

Eigenvalues
        Variance % of var. Cumulative % of var.
dim 1      0.393     2.604                2.604
dim 2      0.376     2.487                5.092
dim 3      0.339     2.248                7.339
dim 4      0.334     2.215                9.554
dim 5      0.326     2.162               11.716
dim 6      0.311     2.062               13.778
dim 7      0.309     2.043               15.821
dim 8      0.299     1.978               17.799
dim 9      0.291     1.929               19.727
dim 10     0.288     1.906               21.634

Cramer's V  0.443    Inertia  15.101
```

4.11.2.2 Interpreting the first axes

We only briefly comment on the results provided by the CA. In the principal plane shown in Figure 4.5 (top), the document cloud does not have a clearly defined shape. It is very dense in the lower part of the central zone, and quite spread out in the upper part, especially in the first quadrant. This already suggests that any obtained partition will not be into visually distinctive clusters.

Figure 4.5 (bottom) shows words in the principal plane limited to those whose contribution is greater than twice the average on at least one of the two axes. Here, proximity between words comes from the fact that they often appear either together in the same answers, or in similar lexical contexts. Therefore, in these plots we observe certain sentence elements close to each other, which helps us to see the main subjects of the free text answers, as well as associations between them, because—generally speaking—answers do not

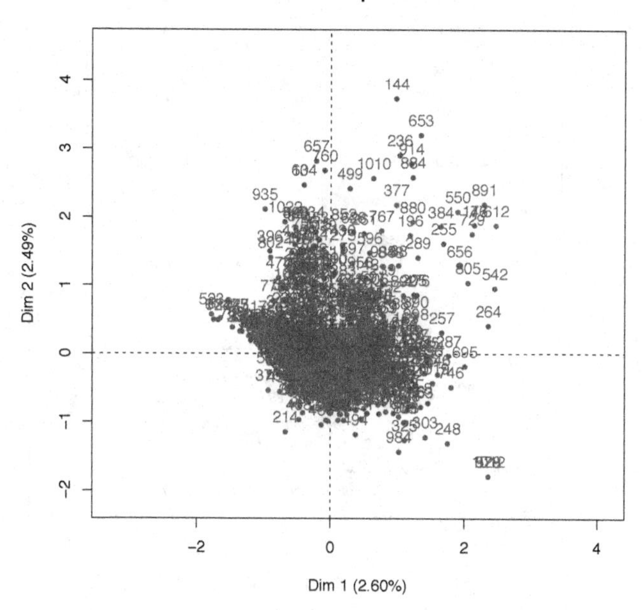

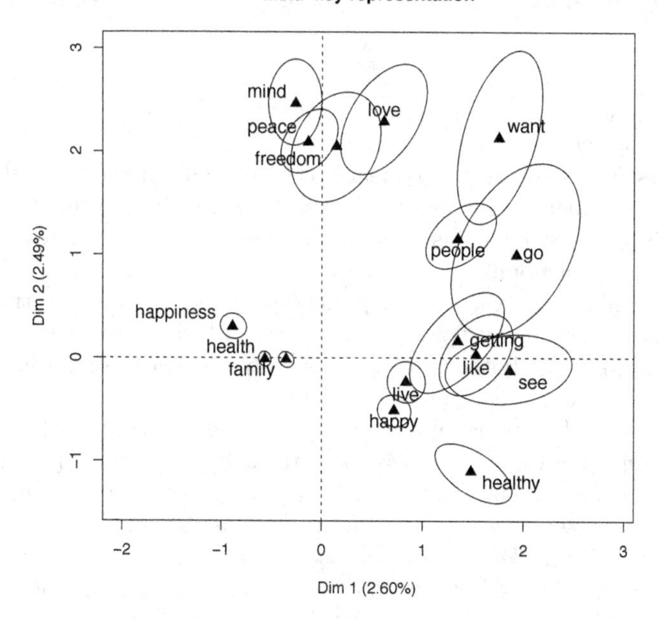

FIGURE 4.5
CA of free text answers: (1,2) plane. Top: the documents; bottom: the highest-contributing words.

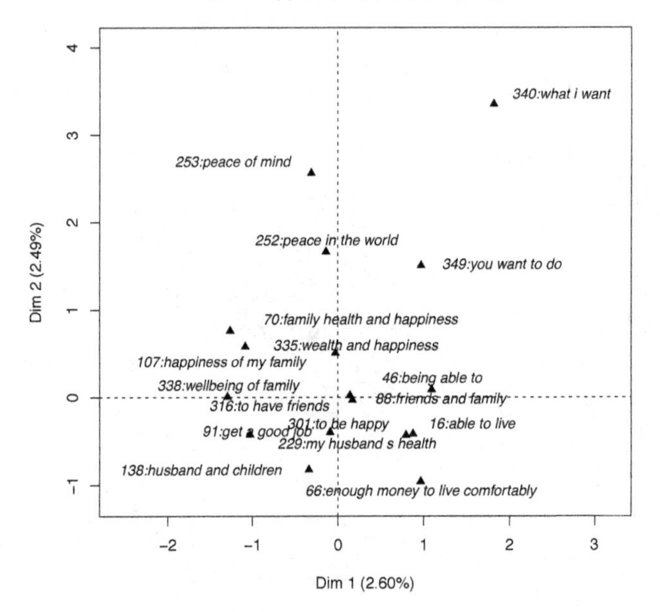

FIGURE 4.6
CA of free text answers: (1,2) plane: supplementary repeated segment representation.

stick to one idea. To give just one example: "*to be happy, healthy, have enough to eat, enough money to live on*".

The first axis contrasts a sequence of verbs and adjectives, visible on the right, with a sequence of nouns to the left. The transition formulas help us to confirm that this corresponds to a division between document responses that use either verbs or nouns.

The second axis, towards the top, highlights a group of words that include *mind*, *peace*, *freedom*, *love*, *want*, and *people*, which correspond to spiritual aspirations. Near the bottom of this axis, only the word *healthy* has a contribution greater than twice the average one.

We see also that *peace* and *mind*, often used together in the expression *peace of mind* (see Figure 4.6), are close to each other on the principal plane in Figure 4.5. They do not have exactly the same coordinates because *peace* is also used on its own as well as in other expressions such as *peace in the world.* Similarly, *family* and *happiness* are close to each other since *my family happiness* is a commonly used expression (see Figure 4.6). In contrast, *happy* and *happiness* are far apart as they are used in different responses and contexts.

This information is also shown in Table 4.5, which can be extended to show meta-keys corresponding to a larger number of axes. This can be useful when trying to choose the number of axes to retain.

TABLE 4.5
An extract of the CA summary: meta-keys corresponding to the first two CA axes. The threshold is set at twice the average word contribution.

```
Words whose contribution is over  2  times the average word contribution

Dimension  1 +
people healthy want like see go happy getting live
Dimension  1 -
health happiness family

Dimension  2 +
peace mind love freedom want people
Dimension  2 -
healthy
```

In this type of data, documents are *anonymous*, identified by a simple label (here, the ID number assigned to each respondent), and contextual data describing the documents should be used (and may be of great value). Contextual data makes it possible to identify—or at least glimpse—features related to contrasts seen on the axes.

To determine statistically significant categories, we can compare their locations on the axes with where they would be if category membership was randomly assigned. The point representing supplementary category k, $\{k = 1,\dots,K\}$ on the axis of rank s has for coordinate value $\sum_{i \in N_{I_k}} \frac{f_{ij}}{f_{\cdot j}} F_s(i)$, where N_{I_k} is the subcloud made up of individuals in category k.

Under the assumption that these individuals could be considered as drawn randomly from the overall sample, this category point should be very close to the cloud's overall CoG. To determine if the distance between it and the overall CoG is large enough to reject the random draw hypothesis, distances are first standardized. In this case, the statistic that is obtained directly follows a standard normal distribution, and can be seen as a test value. Therefore, the random draw assumption can be rejected if the test value is greater than 1.96 or less than -1.96, i.e., in both cases the p-value is less than 0.05. Here, the sign of the test value tells us whether coordinate values judged significantly different to 0 are positive or negative.

Table 4.6 shows the test values of the category points. For instance, the `Age_Educ.>55_Low` point, with a test value of 5.2 on the first axis, corresponds to a point which is significantly different to the CoG on this axis. However, apart from this category, only two of the other `Age_Educ` categories (of which there are nine in total) have significantly different locations to the CoG on this axis (`Age_Educ.31_55_Medium`, with -2.5, and `Age_Educ.>55_Medium`, with -1.97). Of course, we also have to consider this variable as a whole. If we now look at the test values of its categories on the second axis, six are greater than 1.96 or less than −1.96, and furthermore, the categories are arranged in their natural order. These two findings strongly support rejection of the random draw hypothesis. Thus, we can say that in this example, no variable seems to be firmly linked with the first axis, whereas `Education`, considered both alone and with respect to the other variables, is clearly linked to the second.

The scale in Figure 4.7 is stretched compared to the one in Figure 4.5 (top). If the category points were to be transferred to Figure 4.5 (top), they would be very close to the CoG. This is because a certain group of words can be used mainly by one category without implying that all answers in this category can be broken down into these words. This in turn implies that the category point is located much more centrally than the words are, which does not contradict the fact that its location is significantly different from the CoG's location.

TABLE 4.6
Test values for the contextual variables.

```
> round(res.LexCA$quali.sup$v.test[,1:5],3)
```

	Dim 1	Dim 2	Dim 3	Dim 4	Dim 5
Gender.Man	0.020	-0.556	1.998	-2.142	-0.078
Gender.Woman	-0.020	0.556	-1.998	2.142	0.078
Education.E_Low	3.232	-6.013	-2.041	2.917	-2.299
Education.E_Medium	-2.932	1.765	-0.448	-2.547	2.150
Education.E_High	-0.513	6.167	3.574	-0.602	0.273
Age_Educ.<=30_Low	0.500	-1.948	-0.302	-0.078	3.391
Age_Educ.<=30_Medium	-0.116	-1.557	0.189	-5.511	2.061
Age_Educ.<=30_High	1.426	2.964	2.208	-1.863	-0.065
Age_Educ.31_55_Low	-1.536	-4.441	0.179	1.206	-0.227
Age_Educ.31_55_Medium	-2.474	3.246	0.104	0.546	2.119
Age_Educ.31_55_High	-1.223	4.078	3.266	-1.101	-0.701
Age_Educ.>55_Low	5.187	-2.182	-2.503	2.298	-3.529
Age_Educ.>55_Medium	-1.967	1.006	-1.377	2.599	-2.139
Age_Educ.>55_High	-1.533	3.290	-0.398	3.421	1.935
Gender_Educ.M_EduLow	1.486	-5.428	0.152	1.131	-1.886
Gender_Educ.M_EduMed	-0.520	1.948	0.680	-3.572	0.482
Gender_Educ.M_EduHigh	-1.538	4.601	2.633	-0.607	2.116
Gender_Educ.W_EduLow	2.312	-1.829	-2.471	2.281	-0.874
Gender_Educ.W_EduMed	-2.994	0.282	-1.164	0.300	2.102
Gender_Educ.W_EduHigh	0.849	3.785	2.227	-0.210	-1.760

4.11.3 AHC: building the tree and choosing the final partition

The first decision to make is to choose the number of axes to keep for clustering. A closer look at the meta-keys associated with the first ten axes (Table 4.5 only shows those corresponding to the principal plane) turns out to indicate that from the seventh axis on, we find words belonging to earlier axes' meta-

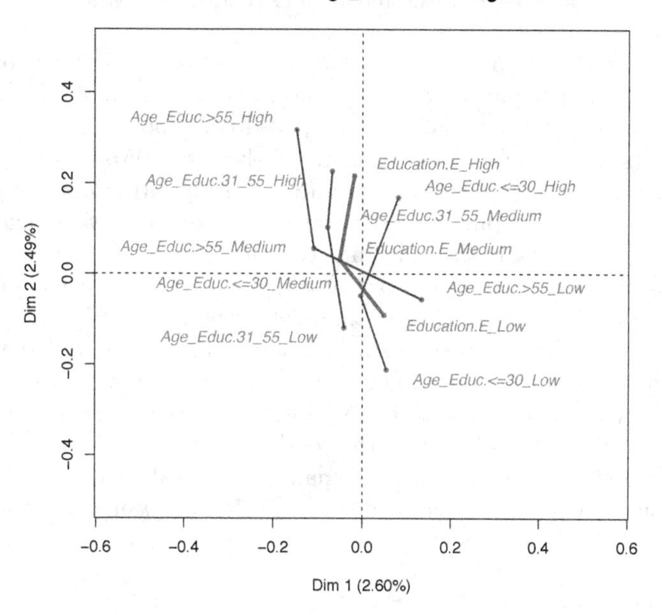

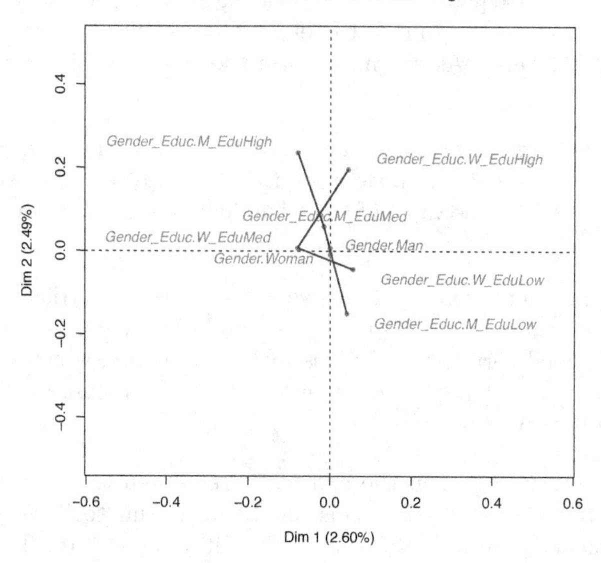

FIGURE 4.7
CA of free text answers: (1,2) plane. Supplementary category representation. Top: `Education` and `Age_Educ` categories; bottom: `Gender` and `Gender_Educ` categories.

keys now found in essentially incoherent meta-keys. We therefore decide to perform clustering based on the individuals' coordinate values on the first six axes.

A plot of the hierarchical tree, shown in Figure 4.8 (top), only brings clarity to the upper part of the tree, due to the large number of documents being clustered. The automatic cut proposed corresponds to a partition into 6 clusters (smallest number of elements: 60, largest: 505).

However, we prefer the partition into a slightly larger number of clusters (i.e., 8) as it avoids keeping the relatively oversized cluster with 505 document responses. Continuing in the same direction, further splits would lead to overly small clusters (less than 30 responses) which would be hard to interpret well alongside a cluster with still more than 300 responses.

The clusters are numbered as a function of the position of their CoG on the first axis. As mentioned earlier, the smallest cluster contains 54 responses (cluster 7, with 5.2% of the total), whereas the largest now contains 379 responses (cluster 1, with 36.5% of the total).

The object `call$t` contains the main output values of the AHC. These concern only the partition obtained before performing k-means iterations after cutting the tree. The most interesting output values are as follows:

- `call$t$within`: the sequence of within-cluster inertias corresponding to partitions into $n = 1, 2, 3, \ldots, I$ clusters. In the $n = 1$ case, i.e., when the tree is complete and all individuals are together in one cluster, the within-cluster inertia is equal to the total inertia of the document cloud, which is in turn equal to the sum of the eigenvalues associated with the retained axes (= 2.081 here). As we move down the tree, the within-cluster inertia decreases.

- `call$t$inert.gain`: this is the gain in between-cluster inertia as we descend the tree. For instance, when we move from $n = 1$ to $n = 2$ clusters, the gain is equal to 0.2058. The value of this gain decreases as we move further down the tree.

- `call$t$quot`: the ratio of successive within-cluster inertias. This is calculated for partitions with between min and max clusters. The automatic cut-off provided corresponds to this ratio's minimum, which is 0.8881 here. Given that the first ratio is calculated for $min = 3$ clusters, this minimum corresponds to 6 clusters.

As mentioned earlier, in the end we have chosen to retain the partition with 8 clusters. Next, the k-means algorithm is run to iteratively reassign each document to the cluster whose CoG it is closest to. The number of reassignments is large, which is to be expected for a fairly continuous cloud like this one (see Figures 4.5 and 4.8) without clearly defined clusters.

For comparison's sake, Figure 4.9 shows the hierarchical tree constructed in the full space, i.e., using the documents' coordinates across all (here = 77)

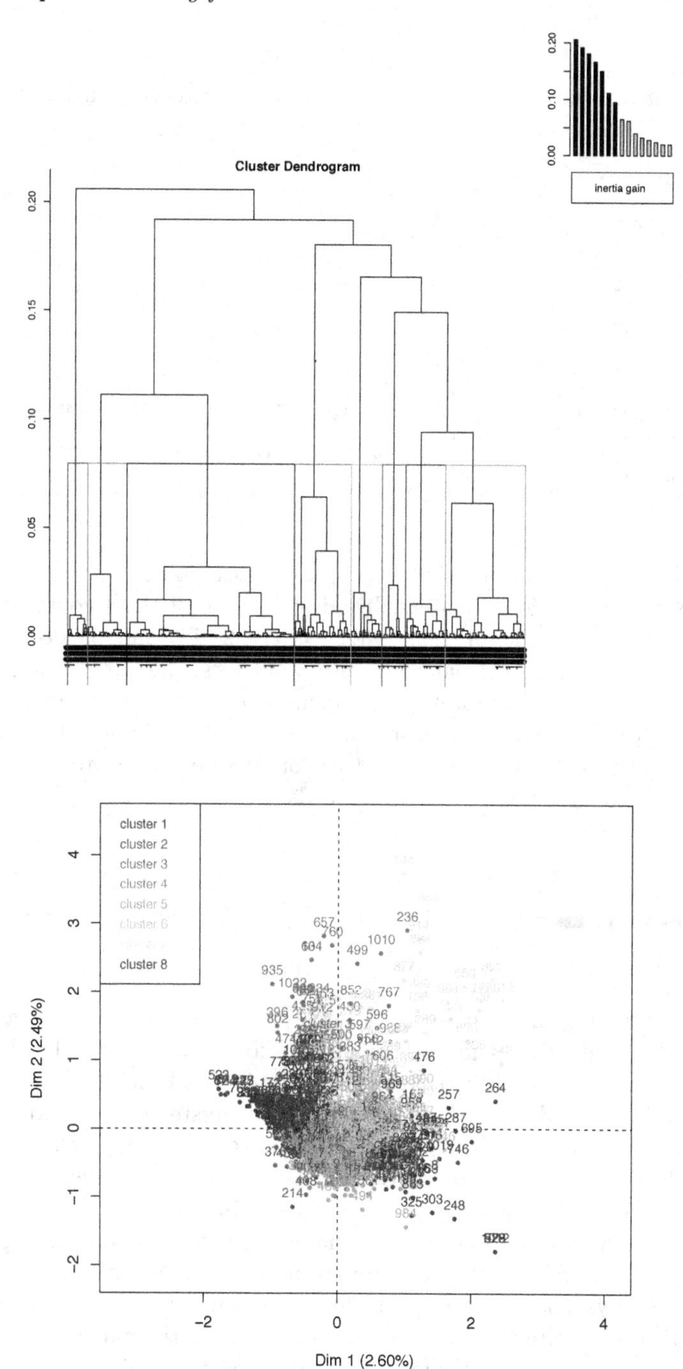

FIGURE 4.8
Clustering the documents. Top: hierarchical tree constructed on the 1039 document responses; bottom: (1,2) CA plane in which document points are colored according to their cluster.

TABLE 4.7
Number of documents in each cluster and outputs describing features of the hierarchical tree.

```
> res.hc$clust.count
          counts
cluster-1    379
cluster-2     55
cluster-3     59
cluster-4     65
cluster-5    208
cluster-6    141
cluster-7     54
cluster-8     78
> round(res.hc$call$t$within[1:10],4)
 [1] 2.0805 1.8747 1.6828 1.5021 1.3361 1.1865 1.0753 0.9808 0.9164 0.8545
> round(res.hc$call$t$inert.gain[1:10],4)
 [1] 0.2058 0.1920 0.1807 0.1660 0.1495 0.1113 0.0944 0.0644 0.0619 0.0396
> round(res.hc$call$t$quot,4)
[1] 0.8976 0.8926 0.8895 0.8881 0.9062 0.9122 0.9343 0.9324
```

factorial axes. The automatically proposed partition is not very interesting because it only has four clusters. In addition, one of them contains 933 documents, while the other three contain only 32, 11, and 63 documents, respectively. If we ask for a partition into 8 clusters, the size of the largest one decreases only a small amount to 883 documents. Unsatisfactory results of this type are often seen when the last axes—essentially residual information—are retained. In fact, these axes often turn out to be related to isolated words rather than to associations between words.

4.12 Describing cluster features

Once partitioning into clusters has been performed, features of each cluster can be described using all of the information available on the documents, whether or not it played an active role in the CA and therefore the clustering. Characteristic words and documents of each cluster can be extracted—see Chapter 5 for more details. As the contextual variables may behave differently across clusters, their relationships with the partition and clusters are studied further if they were selected by the `TextData` function.

Thus, in the case of qualitative variables, we are searching for significantly under- and over-represented categories in clusters. As for quantitative variables, an analysis of variance can identify those which are significantly associated with the partition. This type of description of the partition and clusters provides a large volume of information, only part of which we have space to examine further here.

As seen before, in the case of an aggregate analysis, the categories of the

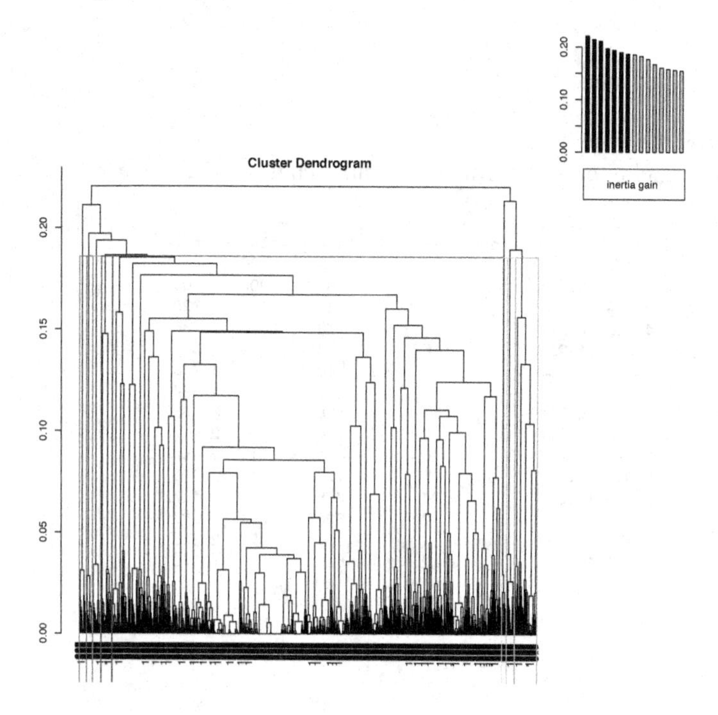

FIGURE 4.9
Hierarchical tree constructed in the full factorial space.

contextual qualitative variables can be recoded in such a way that each of them is considered as a supplementary document. Thus, they are not used to describe the partition and clusters.

4.12.1 Lexical features of clusters

4.12.1.1 Describing clusters in terms of characteristic words

First of all, we have the description of clusters in terms of vocabulary (see an extract of this in Table 4.8). The selection of a cluster's characteristic words will be explained in detail in Chapter 5.

Take cluster 7 as an example. Its CoG occupies a position far to the right of the first axis. Documents in this cluster are generally found in the first quadrant (see Figure 4.8). Remembering that the positive part of the first axis is focused on respondents who favor the use of verbs, we are not surprised to see among the most characteristic words *want*, *go*, *going*, *getting*.

We also note the usage of meta-keys corresponding to the positive part of the first axis, like *love* and *freedom*. On the other hand, the frequent words *family*, *health*, and *happiness* are significantly under-used.

TABLE 4.8
Describing cluster 7 in terms of its most characteristic words.

```
> round(res.hc$desc.wordvar$frequency$'7',3)
          Intern % glob % Intern freq Glob freq  p.value v.test
love        10.277  0.657          26        33    0.000 11.018
want         7.115  0.617          18        31    0.000  8.101
freedom      6.324  0.756          16        38    0.000  6.725
go           4.348  0.378          11        19    0.000  6.236
people       6.324  1.253          16        63    0.000  5.341
music        3.162  0.358           8        18    0.000  4.745
church       2.767  0.318           7        16    0.000  4.384
things       2.767  0.557           7        28    0.001  3.383
going        2.372  0.517           6        26    0.003  2.967
getting      2.372  0.696           6        35    0.015  2.444
happiness    1.976  4.556           5       229    0.044 -2.009
family       9.486 14.027          24       705    0.034 -2.117
good         2.767  6.029           7       303    0.023 -2.267
life         0.791  3.203           2       161    0.021 -2.315
health       4.743 12.177          12       612    0.000 -3.996
```

TABLE 4.9
Describing cluster 7 in terms of its most characteristic documents.

```
res.hc$docspara[[7]]
  DOCUMENT CRITERION      ---------------------TEXT---------------------
1      848 0.8030591      permanent job.my family, my car, having my freedom
                          going out and doing what I want
2      289 0.8100665      peace, no bother, everything going well,
                          (after all when you are getting older you do not
                          want to be upset and messed about). music, my organ
3      880 0.8310451      being happy.being surrounded by people you love,
                          having money, the freedom
                          to do what I want, having this to do
```

```
> res.hc$docsdist[[7]]
  DOCUMENT CRITERION ---------------------TEXT---------------------
1      144  6.135817    become enlightened.love relationship, completing
                        my expectations with society
2      914  4.775583    worship.under that word I would have the 2
                        commandments, (1) to love God above all things then
                        (2) to love thy neighbour as thyself
3      891  4.260221    knowing there are people I can go to if I want to.
                        painting, music, my cat, trees, books
```

4.12.1.2 Describing clusters in terms of characteristic documents

The `LexHCca` function identifies cluster 7's most characteristic documents in two ways. The first criterion, in terms of proximity to the CoG, identifies the documents we call *paragons*. The second, related to the distance to the other CoGs, selects what we call *specific* documents. The literal texts of these documents for the two criteria are found respectively in the `$docspara` and `$docsdist` objects output by the `LexHCca` function. In Table 4.9 we show the three most characteristic documents according to the two criteria. A comprehensive analysis of the paragons and specific documents of cluster 7 can provide more context on the cluster's most characteristic words, along with details on their semantics. Indeed, in these documents we find expressions like: *my family, my car, having my freedom, going out and doing what I want,* as well as: *being happy, being surrounded by people you love, having money, the freedom to do what I want, having this to do.* Note also that very common words like *family* and *money* can be found in the characteristic responses without actually being characteristic words of the cluster. In the case of more outlying clusters, their specific documents are, by construction, more characteristic of the cluster, since they are the ones furthest from the other CoGs.

4.12.2 Describing clusters using contextual variables

Describing the partition and its associated clusters using contextual variables (see Table 4.10) leads in our case to a socio-economic description of the clusters.

4.12.2.1 Describing clusters using contextual qualitative variables

First, we use a χ^2 test to evaluate the strength of the link between the partition (a qualitative variable whose categories are the clusters as identified by their number) and each of the additional qualitative variables. Calculations are performed separately for each of them.

These are then ordered according to the value of the p-values associated with the χ^2 tests, from smallest to largest. Indeed, it makes sense to consider that the more the independence hypothesis is rejected, the more a variable is associated with the given partition.

In our example, the four variables considered are associated with the partition (p-values below 0.05), with `Age_Educ` showing the strongest link.

We can extend our initial analysis on the relationship between the partition and each of the qualitative variables by looking for categories that are significantly over- or under-represented in each cluster.

As an example, let us run through the relevant calculations for cluster 7 and the `Woman` category of the `Gender` variable. We will say that this category, denoted k, whose number of members in cluster q is given by I_{qk}, is over-represented in this cluster, which has I_q members in total, if the proportion of women in it ($= I_{qk}/I_q$) is significantly superior to the proportion of women

TABLE 4.10
Description of the partition and cluster 7 in terms of contextual variables.

```
> res.hc$desc.wordvar$test.chi2

$test.chi2
                  p.value df
Age_Educ     1.567289e-119 56
Gender_Educ   1.685412e-50 35
Education     1.336369e-38 14

> round(res.hc$desc.word$category$'7',3)
                        Cla/Mod Mod/Cla Global p.value v.test
Gender_Educ=W_EduHigh    14.697  20.158  6.904   0.000  7.139
Age_Educ=<=30_High       15.464  17.787  5.790   0.000  6.925
Education=E_High          9.429  26.087 13.928   0.000  5.250
Age_Educ=31_55_Medium     6.946  22.925 16.614   0.008  2.661
Gender=Woman              5.712  61.660 54.337   0.016  2.406
Age_Educ=>55_High         0.952   0.395  2.089   0.032 -2.144
Gender=Man                4.227  38.340 45.663   0.016 -2.406
Age_Educ=<=30_Low         0.000   0.000  1.651   0.013 -2.477
Gender_Educ=W_EduLow      3.641  18.577 25.686   0.006 -2.725
Gender_Educ=M_EduLow      2.882  11.858 20.712   0.000 -3.762
Education=E_Low           3.302  30.435 46.399   0.000 -5.295
Age_Educ=31_55_Low        2.011   8.696 21.767   0.000 -5.645
```

overall ($= I_k/I$). Here, 54.34% of the overall sample are women, as opposed to 61.66% in cluster 7. Under the independence hypothesis, these two proportions should be close. To calculate the p-value, and thus decide on the significance of the over-representation of women in cluster 7, we proceed as follows. We assume the null hypothesis, which means considering that the I_q respondents in cluster q were randomly drawn without replacement from the overall set of I respondents. The random variable whose observed value is I_{qk}, equal to the number of respondents who are in the Woman category in cluster q, follows a hypergeometric distribution with parameters I, I_k, and I_q.

We can therefore calculate the probability under the null hypothesis to observe a value greater than or equal to the observed value. This probability, calculated as:

$$P(q,k) = \sum_{x=I_{qk}}^{I_k} \frac{\binom{I_k}{x}\binom{I-I_k}{I_q-x}}{\binom{I}{I_q}},$$

directly corresponds to the p-value associated with the test. As is usual, if this value is small (generally, less than 5%), we say that category k is significantly over-represented in cluster q. This calculation is performed for each category. The most over-represented category is that which corresponds to the smallest p-value. Significantly over-represented categories (those associated with a p-value below 0.05) from all of the qualitative variables pooled together can then be sorted from smallest to largest p-value, or equivalently, from largest to smallest test value: the smaller the p-value, the more positive the corresponding test value is.

In the same way, we can look for significantly under-represented categories. We say that a category k with I_{qk} elements in cluster q is under-represented in this cluster, which has a total of I_q elements, if the proportion of documents with category k in it $(= I_{qk}/I_q)$ is significantly less than the proportion I_k/I of documents with category k in the overall group of I respondents. To calculate the p-value and thus conclude as to the significance of the under-representation, we follow the same reasoning as before, under the null hypothesis. The random variable whose observed value is equal to I_{qk} follows a hypergeometric distribution with parameters I, I_k and I_q. We can therefore calculate the probability under the null hypothesis to observe a value equal to or less than the observed value. This probability, calculated as:

$$P(q,k) = \sum_{x=1}^{I_{qk}} \frac{\binom{I_k}{x}\binom{I-I_k}{I_q-x}}{\binom{I}{I_q}},$$

directly corresponds to the p-value associated with the test. It is again useful to convert this into a test value. The smaller the p-value, the more negative the corresponding test value is.

Hence, cluster 7 is characterized by an over-representation of the `Woman` category (and thus an under-representation of `Man` since the `Gender` variable has only two categories). Overall, we can say that this cluster attracts, in particular, the young with high levels of education, especially women.

4.12.2.2 Describing clusters using quantitative contextual variables

For each of the quantitative variables, we can calculate the squared correlation ratio between it and the partition:

$$\eta^2 = \frac{\textit{between-cluster inertia}}{\textit{total inertia}}.$$

This calculation also corresponds to that of the R^2 coefficient in analysis of variance. Note further that we can also see the description of a partition as a mean comparison test, whose null hypothesis is the equality of the mean in each cluster, and whose p-value is like the one in analysis of variance.

Here also we can enlarge on our judgment of the relationship between the partition and each of the quantitative variables by comparing the mean of each cluster with the overall mean. For each cluster and quantitative variable—which we denote X—we calculate:

$$\frac{\overline{x_q} - \overline{x}}{\sqrt{\frac{s^2}{I_q} \cdot \frac{I-I_q}{I-1}}},$$

where $\overline{x}$ is the mean of X in the sample, $\overline{x_q}$ its mean in cluster q, s^2 its variance in the sample, I_q the number of elements in q, and I the total number in the sample.

The null hypothesis is: the values taken by X for individuals in cluster

q are drawn randomly without replacement from the values taken by X on the whole sample. Under this hypothesis, the expectation and variance of the random variable $\overline{X_q}$, the mean for individuals in cluster q, are given by:

$$E(\overline{X_q}) = \overline{x} \quad \text{and} \quad V(\overline{X_q}) = \frac{s^2}{I_q} \times \frac{I-I_q}{I-1}.$$

If in the population the variable X is normally distributed, then under the null hypothesis, $\overline{X_q}$ is too. If X is not normally distributed, we can nevertheless suppose that the distribution of $\overline{X_q}$ does not stray far from normality.

In this case, the proposed statistic corresponds to the standard deviation between the mean for individuals in the cluster and the overall mean, and thus to the test value, from which the p-value is easily deduced. Either the p-value or the test value, equivalently, can be used to reject or not reject the null hypothesis above, which can be restated as: the mean of X in the cluster is equal to the overall mean.

4.12.3 Implementation with Xplortext

The code in `script.chap4.b.R` performs the following tasks:

- The `TextData` function builds the LT. A group of tool words, provided by the user, are removed. The contextual variables of interest are stated. Only words found in at least 15 source documents (= responses) are retained.

- The `LexCA` function runs CA on the LT. No plots are asked for. Words and documents are selected if their contribution is at least twice the average contribution.

- The `plotLexCA` function plots in succession the documents (= individual responses), meta-keys and their confidence ellipses, supplementary categories and perhaps their trajectories, and last, the supplementary quantitative variable.

- The `LexCA` function, asking that 6 axes be retained, followed by the `LexHCca` function, together lead to hierarchical clustering without constraints being run, using coordinate values from the first 6 axes.

- The object created by the `LexHCca` function gives a detailed description of the partition and the clusters.

- The `LexCA` function, retaining all axes, followed by the `LexHCca` function, allow us to see what the clustering results would look like if we kept all of the axes.

4.13 Summary of the use of AHC on factorial coordinates coming from CA

In this chapter, we have focused on AHC, which is the most commonly-used clustering method associated with CA. We have presented several examples showing the type of results provided by the method when associated with CA. Documents are clustered according to their coordinate values on the retained factorial axes. All proximities between documents are defined in this space.

AHC can help reveal a hierarchical structure in a corpus, which is particularly interesting in the case of chronological corpora. The use of Ward's variance-based aggregation criterion is in tune with CA, which also decomposes the inertia. Nevertheless, this criterion may cause inversions when running contiguity-constrained clustering, so it is preferable to use the complete-linkage method here.

The main points to remember are the following:

- Clustering makes it possible to summarize similarities in large numbers of documents, grouping them as a small number of homogeneous clusters based on their lexical profiles.

- The lexical profile of each cluster can be calculated; it is equal to the weighted average profile of the documents in it.

- After AHC has been performed, we end up with a nested sequence of partitions, from one extreme: each document constitutes a cluster, to the other: there is a single cluster with all documents in it. This sequence of nested partitions is represented as a hierarchical tree.

- Progression in the value of the aggregation criterion is displayed as a bar chart, which makes it possible to choose a cut in the tree corresponding to a large jump between successive values of it.

- Cutting the tree at a certain level means choosing a partition with as many clusters as there are branches of the tree that were cut through.

- Clusters can be highlighted on the factorial plane, thus making it possible to visualize the distances between them and the ways in which they deviate from the global CoG.

- Each cluster can then be described in terms of the words it significantly over- or under-uses (see Chapter 5 for more details). These words are the ones that contribute most to the distance of a cluster from the CoG.

- Each cluster can also be described in terms of contextual information that significantly differentiates it from the overall set of documents.

5

Lexical characterization of parts of a corpus

Results from CA and clustering, the multidimensional methods presented in Chapters 3 and 4, can be built upon by also identifying which words characterize each document, as opposed to the corpus as a whole. This task is based on one-dimensional calculations, performed successively on each word. Documents need to be long enough that such comparisons make sense. In this context, the search for characteristic words will often be applied to aggregate groups of documents formed with respect to various criteria, depending on the nature of the corpus and the objectives of the analysis.

When the documents included in these aggregate groups are sufficiently short, we can also identify those of them which are most representative of the aggregate, calling them *characteristic documents.* In concrete terms, this chapter is devoted to selection criteria for characteristic words and characteristic documents.

Corpora of free text answers lend themselves particularly to this approach. Free text answers, i.e., source documents (which are usually short), can be aggregated according to either categories of a qualitative variable or clusters.

Each aggregate document thus formed can then be characterized by significantly over- and/or under-used words, in comparison with their use in the whole corpus. In addition, we can also try to identify a small number of documents which seem to best represent an aggregate.

If the analysis concerns a corpus of long documents, such as all of an author's works or politician's press conferences, then we will only be interested in characteristic words. In the case of a chronological corpus, we will also see that the lexical characterization of the nodes of the tree can illustrate the hierarchical structure of the corpus (see Section 5.3.3). More generally, we can examine a corpus that has been divided into parts, no matter how these parts were formed, or whether each is composed of a single document or more. Each part can then be described in terms of the significantly over- and/or under-used words in it. If a part is composed of many short documents, it may make sense to look for the whole documents that are the most characteristic of it.

5.1 Characteristic words

The relevant table is now a "parts of the corpus" $\times$ words table, noted $\mathbf{Y}$. Its y_{qj} entry gives the number of times part q uses word j. Correspondingly, $y_{.j}$ is the frequency of word j in the whole corpus, and $y_{q.}$ the size (number of occurrences) of part q. N is the overall size of the corpus. This table is turned into a relative frequency table, denoted $\mathbf{F}$, by dividing each term by N.

The association rate τ_{qj} (see Section 2.2.4) here measures the strength of association between a part and a word:

$$\tau_{qj} = \frac{N \times f_{qj}}{N \times f_{q.} \times f_{.j}} = \frac{f_{qj}}{f_{q.} \times f_{.j}}.$$

If $\tau_{qj} > 1$, the word j is over-used in part q compared to the rest of the corpus; correspondingly, $\tau_{qj} < 1$ represents under-use.

To conclude as to under- or over-use of a word j in part q, we run the same test we used when looking at under- or over-use of a category k in a cluster q (see Section 4.12.2.1). The word j has a total frequency of $y_{.j}$ and part q is made up of $y_{q.}$ items. Essentially, we have to compare the observed frequency y_{qj} of the word j in part q with the expected frequency if this part had been made up of $y_{q.}$ occurrences chosen at random from a set of size N. Then, following the same reasoning as before, we know that the relevant random variable X, whose observed value is y_{qj} (the frequency of word j in part q), follows a hypergeometric distribution with parameters N, $y_{.j}$, and $y_{q.}$, thus allowing us to directly calculate p-values.

Over-represented words in a certain part of a corpus are called *positive characteristic elements*, while under-represented ones are called *negative characteristic elements.* To make the results easier to read, aside from the p-values themselves, we also provide the corresponding test values (see Section 4.12.2). For example, a p-value below 5% corresponds to a test value that is either less than -1.96 (significant under-use) or greater than 1.96 (significant over-use). Words not characterizing any part at all are simply called *banal* words.

5.2 Characteristic words and CA

The more a word is under- or over-used in a part q, the more it contributes to the χ^2 distance between the part and the CoG of the document cloud, which is also the CoG of the part cloud. In effect, this distance is expressed as follows:

$$d^2(q, CoG) = \sum_{j=1}^{J} \frac{1}{f_{.j}} \left(\frac{f_{qj}}{f_{q.}} - f_{.j} \right)^2,$$

and thus the contribution of word j to this distance can be written as a function of τ_{qj}:

$$\frac{1}{f_{.j}}\left(\frac{f_{qj}}{f_{q.}} - f_{.j}\right)^2 = \frac{\left(f_{qj} - f_{q.}f_{.j}\right)^2}{\left(f_{q.}^2 f_{.j}\right)} = f_{.j}\left(\tau_{qj} - 1\right)^2.$$

It is of interest to try to relate characteristic words to CA results. On a plane, a part q of a corpus is strongly attracted by its positive characteristic elements and strongly repulsed by its negative characteristic elements. Thus, it is useful to highlight the positive characteristic elements on the factorial planes and link them to the parts they characterize well. The same can be done with the negative characteristic elements, though their role, which is to repel the parts that under-use them, is less visually obvious on plots.

It should be kept in mind that characteristic words of a part are only so in relation to the whole corpus, and do not constitute a good tool for comparing the parts among themselves and describing their similarities. Two parts with similar lexical profiles as indicated by their proximity in CA planes may or may not share some characteristic words.

The results of the selection of characteristic words and those from CA correspond to different points of view: a characterization of the parts with respect to the CoG, versus a summary of a pairwise comparison of the parts; these two points of view complement each other.

5.3 Characteristic words and clustering

Two very different cases must be considered here. In the first, clustering is done using the verbal content of documents, i.e., using their factorial coordinates output from a CA of the documents × words table to obtain homogeneous document clusters with respect to verbal content. In the second, documents are clustered in terms of their values of the contextual variables, in order to obtain homogeneous clusters with respect to the values of these variables.

5.3.1 Clustering based on verbal content

When clustering is based on verbal content, words play an active role. The aim is then to identify different words' influence in the construction of clusters. Over-used words in a cluster are strongly shared by documents in it, while under-used onés are usually avoided. The most characteristic words are the ones most responsible for the formation of the clusters. They are therefore identified *a posteriori*. This active role of words often leads to finding many characteristic words with small p-values. Of course, the p-values tend to be much greater for words playing an illustrative role.

5.3.2 Clustering based on contextual variables

We assume here that documents are free text answers, but the following can also be applied profitably when dealing with many short documents described by a large number of contextual variables. We have seen that CA applied to an ALT created from a mixed contextual variable like `Age_Educ` makes it possible to study the relationship between the textual content of answers and age group for a fixed level of education, and/or education for a fixed age group. We would now like to be able to consider more than two variables at once (e.g., *age, gender, level of education* and *urban/rural habitat*) in order to compare the aggregate documents of groups of respondents, each of which is homogeneous from a socio-demographic point of view.

However, an analysis involving all combinations of the categories with three or more variables selected is impractical, and further, some of these would be empty or involve a very small number of items. The method we use instead to combine variables is to run a multiple correspondence analysis (MCA) based on them, followed by clustering the documents in terms of their coordinate values on the factorial axes. Document clusters which are relatively similar in terms of the variables used in the MCA can then be obtained. This type of partition is called a *working demographic partition* by Ludovic Lebart. Then, it is a question of looking for potential links between the partition and the verbal content of the free text answers by grouping together those found in the same cluster. We can then look for each cluster's characteristic words and documents. To calculate these, we form the document aggregating the set of free text answers that belong to the cluster. There is a strong relationship if there are many words that characterize the clusters.

The absence of a relationship would instead correspond to a small number of words that characterize the clusters. Surveys including free text answers are a prime target for this kind of technique, given the numerous contextual variables they collect.

5.3.3 Hierarchical words

Chronological corpora, whose temporal structure can be uncovered and then visualized as hierarchical trees with contiguity constraints, require a specific approach. Here, we are looking to describe this structure in terms of the lexical features that characterize it. All nodes, whether terminal or intermediate, group together contiguous sequences of the corpus. Some words best characterize short sequences, others longer ones. This observation leads us to associate hierarchical trees with contiguity constraints on the documents with what we call *hierarchical words*.

First, the characteristic words are computed for every node of the hierarchy, including the terminal ones that correspond to the parts. Then, each word is assigned to the node for which it presents the smallest p-value (or largest test value). To better understand the information they provide, these hierarchical words are then shown on the dendrogram of the hierarchical tree. The current version of **Xplortext** does not allow for an automatic visual output of hierarchical words, so they must be added manually. An example of this is shown in Section 7.4.

5.4 Characteristic documents

In the case where parts are formed by grouping together short documents, identifying the most "characteristic" documents of each part can help to put the characteristic words in their proper context. These documents, known as *characteristic documents*, can be identified based on the characteristic words they contain. In practice, they are identified using an empirical rule. The rule, which has the advantage of being simple, is to associate with each document the average test value of the words it contains. This will be the criterion we use. The presence of positive characteristic elements in a document raises the average, while that of negative characteristic elements lowers it. The value of the criterion is calculated for all of the documents in the part. Then, the documents are ordered based on decreasing values, from most to least characteristic. We then keep only a predefined number of documents, those associated with the highest values of the criterion, so as to retain a variety of lexical features of the part being examined. Note that this criterion favors short documents—those in which positive characteristic elements are concentrated.

5.5 Example: characteristic elements and CA

This example comes from *Life_UK*, already used in earlier chapters. Results are output from the `LexChar` and `LexCA` functions of **Xplortext**. The individuals' free text answers are here regrouped in terms of the six categories of the variable *Gender_Age*: *men/women 30 or less* (134 and 132 respondents), *men/women 31-55 years* (194 and 251 respondents), and *men/women over 55* (167 and 162 respondents).

5.5.1 Characteristic words for the categories

The characteristic words found for the youngest and oldest categories of both genders are shown in Table 5.1 (for men) and Table 5.2 (for women). The columns of these tables show, successively, the word in question, the percentage of occurrences which correspond to this word in each part, the percentage of occurrences which correspond to this word in the whole of the corpus, the frequency of the word in each part, the frequency of the word in the whole corpus, the p-value, and the test value.

For a given part, only the words associated with a p-value less than the one indicated in the `proba` argument of the `LexChar` function are shown, ordered in terms of decreasing test value. The default value (which we use here) for the `proba` argument is 0.05.

TABLE 5.1
Characteristic words of the youngest and oldest male age groups.

Group1: M<=30

Word	Intern %	glob %	Intern freq	Glob freq	p.value	v.test
work	4.726	2.212	31	118	5e-05	4.077224
job	5.030	2.681	33	143	4e-04	3.538651
friends	3.811	2.175	25	116	0.00636	2.728765
social	0.915	0.281	6	15	0.01269	2.492345
nothing	2.439	1.331	16	71	0.02164	2.296548
future	0.915	0.319	6	17	0.02530	2.236784
else	2.134	1.162	14	62	0.03254	2.137661
things	1.220	0.525	8	28	0.03291	2.133192
happy	1.372	2.568	9	137	0.04045	-2.04917
grandchildren	0.000	0.562	0	30	0.03857	-2.068736
children	1.220	2.456	8	131	0.02874	-2.187093
health	7.165	11.474	47	612	0.00013	-3.821116
husband	0.000	1.800	0	96	1e-05	-4.526958

Group3: M>55

Word	Intern %	glob %	Intern freq	Glob freq	p.value	v.test
wife	3.517	1.425	30	76	0.00000	4.849302
good	8.089	5.681	69	303	0.00189	3.106489
grandchildren	1.407	0.562	12	30	0.00272	2.997513
life	4.572	3.018	39	161	0.00788	2.657382
help	0.938	0.375	8	20	0.01730	2.380332
daughter	0.938	0.394	8	21	0.02411	2.255442
go	0.821	0.356	7	19	0.04510	2.003765
anything	0.821	0.356	7	19	0.04510	2.003765
security	0.117	0.750	1	40	0.01586	-2.412096
children	1.290	2.456	11	131	0.01537	-2.423472
friends	1.055	2.175	9	116	0.01311	-2.480867
job	1.407	2.681	12	143	0.01072	-2.551844
husband	0.117	1.800	1	96	0.00000	-4.773347

Table 5.1 and Table 5.2 show that the word *job* is a positive characteristic element in the case of the youngest male and female age groups. It is also a positive characteristic element for the 31-55 years male age group, whose characteristic words are not reproduced here. All of the other categories significantly under-use *job*. The *men from 31 to 55* and *men over 55* categories over-use *wife*, while all categories of women over-use *husband*, which—though

TABLE 5.2
Characteristic words of the youngest and oldest female age groups.

CHARACTERISTIC WORDS
(DETAILED INFORMATION)

Group4: W<=30

Word	Intern %	glob %	Intern freq	Glob freq	p.value	v.test
happy	5.207	2.568	34	137	6e-05	4.006928
job	5.207	2.681	34	143	0.00016	3.779284
money	5.054	3.225	33	172	0.01031	2.565186
husband	3.216	1.800	21	96	0.01041	2.561753
grandchildren	0.000	0.562	0	30	0.03932	-2.060799
wife	0.153	1.425	1	76	0.00107	-3.271703

Group6: W>55

Word	Intern %	glob %	Intern freq	Glob freq	p.value	v.test
husband	3.721	1.800	31	96	6e-05	4.004715
friends	3.601	2.175	30	116	0.00551	2.775835
health	14.286	11.474	119	612	0.00796	2.653732
food	0.960	0.431	8	23	0.03784	2.076636
wife	0.600	1.425	5	76	0.02915	-2.181455
freedom	0.120	0.712	1	38	0.02489	-2.24303
work	0.840	2.212	7	118	0.00207	-3.080207
job	0.360	2.681	3	143	0.00000	-5.255352

not necessarily surprising—was not seen in the other languages in which the survey was done. What is more surprising is that *women over 55* share with *men 30 and under*—and only with them—the characteristic word *friends*. It is also perhaps surprising that *women over 55* do not over-use the word *grandchildren*, while men of the same age do. Lastly, among the five most-used nouns (*family, health, happiness, money* and *life*), only *health*, *money* and *life* are either positive or negative characteristic elements. Indeed, *health* negatively characterizes *men 30 and under* and positively characterizes *women over 55*, while *money* positively characterizes *women 30 and under* and *life* positively characterizes *men over 55*. This said, the most frequently used words are banal ones whose usage does not differ between categories. Based on these characteristic words, it is clear that what is considered important in life varies with age. Nevertheless, we have to keep in mind that the characteristic words of a category are those which differentiate it from the rest of the sample. A consequence of this is that two categories can have rather close lexical profiles without sharing any characteristic words. We also note that a smooth change in vocabulary usage with age can only be seen clearly on the CA planes—see the following section for more details.

5.5.2 Characteristic words and factorial planes

In order to place the positive characteristic elements on the principal factorial plane, we run a CA of the *Gender_Age* × *words* table. The total inertia, equal to 0.17, and Cramér's V , equal to 0.18, have higher values than when applying CA on the *Gender_Educ* × *words* table (0.10 and 0.14, respectively, see Section 3.2.3). Here, the inertia of the two tables can be directly compared as the tables are the same size, while the Cramér's V values do not bring much to the discussion. The categories are shown on the principal factorial plane in Figure 5.1. On this same plane, but on small separate plots, we also show the youngest and oldest age groups of men and women, along with the associated positive characteristic elements. We can see that the male categories have more characteristic words (8 each) than the female ones (4 each). The characteristic words for the oldest male age group are located in the top-left quadrant, while those of the youngest male age group are found in the positive half of the first axis, in both quadrants. The trajectory for the women's categories is shorter than that of the men's, which is related to the smaller number of characteristic words for women. For both categories of women shown, these are found in the negative part of the second axis, but not in a way that is specific to each category.

5.5.3 Documents that characterize categories

Determining the characteristic documents of a part is meaningful, as long as that part is an aggregate of short individual documents. The `LexChar` function outputs the labels of the five most representative responses of each part (here, part = free text answers grouped by category) followed by their text.

The criterion calculated for each document(=free text answer, here), is the mean of the test values associated with its words for the part to which the document belongs; only the retained words are taken into account. These values are rather small (see Table 5.3).

5.6 Characteristic words in addition to clustering

An example showing the use of characteristic words to describe clusters was given in Section 4.11. Table 5.4 shows detailed results from that example, corresponding to the largest (cluster 1) and smallest (cluster 7) clusters. We see in particular that the sets of words characterizing the clusters are particularly bountiful and/or associated with particularly large test values here. This is

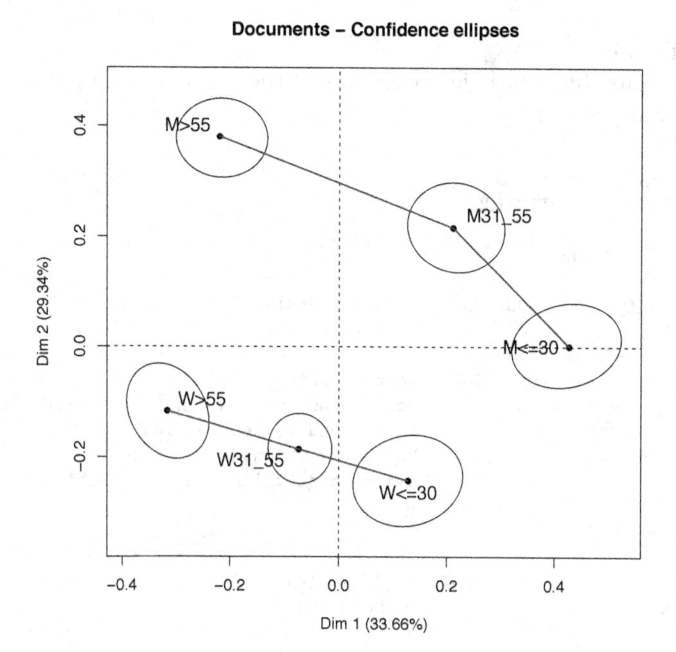

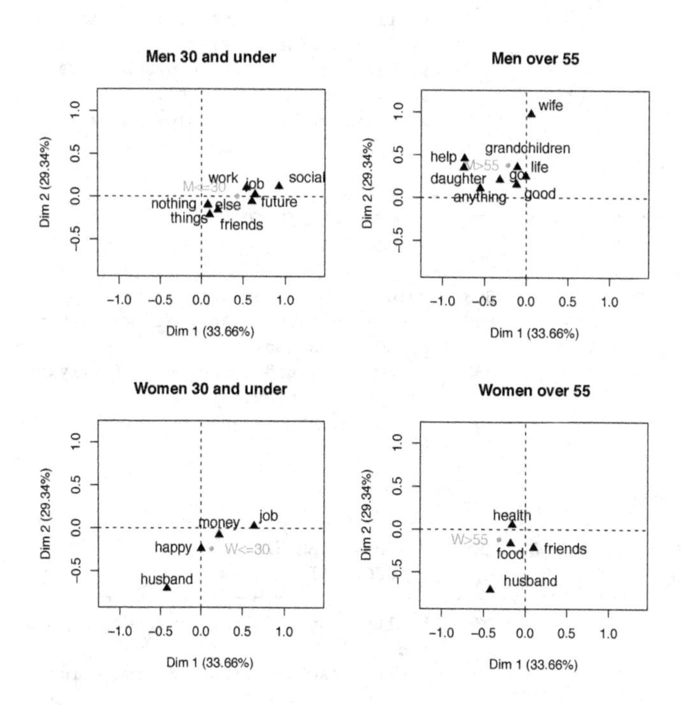

FIGURE 5.1
CA of the *Gender_Age* × words table: (1,2) plane. Top: all of the categories. Bottom: the most-separated categories (for men and women) and their associated positive characteristic elements.

TABLE 5.3
The first four characteristic documents of the youngest and oldest age groups for men and women.

CHARACTERISTIC DOCUMENTS
(WORDS FREQUENCY CRITERION)

GROUP 1: M<=30

CLASSIFICATION CRITERION	DOCUMENT	CHARACTERISTIC DOCUMENT
4.077 ----	655	work. fishing
4.077 ----	872	work. studying and achieving something in the end to become rich
4.077 ----	1041	honest. work hard, honesty
2.729 ----	521	a job. most things, a job, nothing else

GROUP 3: M>55

CLASSIFICATION CRITERION	DOCUMENT	CHARACTERISTIC DOCUMENT
2.652 ----	1003	good health. my wife
2.510 ----	593	to live till I am until 100, and the wife of course.my wife (...)
2.425 ----	27	my wife, she gave me courage to carry on even in the bad times.(...)
2.425 ----	263	my wife. books and music, garden

GROUP 4: W<=30

CLASSIFICATION CRITERION	DOCUMENT	CHARACTERISTIC DOCUMENT
4.007 ----	608	being happy. everybody being happy
3.286 ----	886	making my father happy and my mother proud. more money
2.671 ----	830	being happy and making sure family was happy. companionship
2.588 ----	700	my boyfriend. my job, money, just being happy

GROUP 6: W>55

CLASSIFICATION CRITERION	DOCUMENT	CHARACTERISTIC DOCUMENT
2.715 ----	1037	health. coping alone, friends, garden and birds
2.654 ----	28	health. law and order, rising prices
2.654 ----	659	health.
2.359 ----	243	husband. family, health, socialising, friends

due to the fact that the AHC was run in terms of the verbal content of responses, starting from a previous CA of the documents × words table. Words thus play an active role, and their power for characterization purposes is, by construction, artificially high. We also note that cluster 1, the largest with 379 respondents, is characterized by—among other things—the most common nouns. To keep the size of Table 5.4 manageable, the `proba` argument in the `LexChar` function was chosen small—here 0.001.

5.7 Implementation with Xplortext

The code in `script.chap5.R` performs the following tasks:

- The `TextData` function builds the ALT by aggregating responses according to categories of the `Gender_Age` variable. The corpus corresponds to columns 9 and 10. Only words found in at least 15 documents (source documents) are retained. The stoplist proposed in the **tm** package is used to remove tool words.
- The `LexChar` function selects the characteristic words and documents (characteristic source documents) of each of the six document categories.
- The `LexCA` function runs a CA of the ALT in order to plot the characteristic words on the factorial planes.
- The `plot.LexCA` function outputs plots of the characteristic words of the youngest and oldest age groups (for men and women) on the principal factorial plane of the CA above.

The final three points show how the results from Section 5.6 can be obtained:

- The `TextData` function constructs the LT. The corpus corresponds to columns 9 and 10. Only the words found in at least 15 documents (source documents, which are also the documents being analyzed here) are retained. A list of tool words provided by the user are removed.
- The `LexCA` function performs a CA of the LT.
- The `LexHCca` function regroups the non-aggregate documents into 8 clusters. Then, clusters 1 and 7 are described (as an example) in terms of their characteristic words.

TABLE 5.4
Characteristic words for clusters 1 and 7 of the 8-cluster partition (see Section 4.11.3).

```
> round(res.hc$desc.wordvar$'1',3)

             Intern % glob % Intern freq Glob freq  p.value v.test
happiness      10.710  4.556         166       229   0.000 13.222
health         20.323 12.177         315       612   0.000 11.339
family         20.581 14.027         319       705   0.000  8.656
security        2.000  0.796          31        40   0.000  5.943
contentment     1.484  0.617          23        31   0.000  4.801
good            8.194  6.029         127       303   0.000  4.143
wellbeing       1.032  0.418          16        21   0.000  4.068
satisfaction    1.032  0.438          16        22   0.000  3.838
friends         3.484  2.308          54       116   0.000  3.496
keep            0.323  0.995           5        50   0.001 -3.318
daughter        0.000  0.418           0        21   0.001 -3.336
peace           0.710  1.572          11        79   0.001 -3.361
well            0.194  0.856           3        43   0.000 -3.598
see             0.000  0.517           0        26   0.000 -3.820
husband         0.839  1.910          13        96   0.000 -3.839
able            0.258  1.114           4        56   0.000 -4.143
want            0.000  0.617           0        31   0.000 -4.256
time            0.000  0.617           0        31   0.000 -4.256
healthy         0.129  0.915           2        46   0.000 -4.285
standard        0.000  0.657           0        33   0.000 -4.419
happy           1.226  2.726          19       137   0.000 -4.558
people          0.258  1.253           4        63   0.000 -4.611
just            0.129  1.035           2        52   0.000 -4.710
mind            0.065  0.935           1        47   0.000 -4.856
children        1.032  2.606          16       131   0.000 -4.944
like            0.000  0.796           0        40   0.000 -4.952
enough          0.194  1.353           3        68   0.000 -5.311
wife            0.258  1.512           4        76   0.000 -5.404
living          0.129  1.353           2        68   0.000 -5.721
live            0.129  1.671           2        84   0.000 -6.606

> round(res.hc$desc.wordvar$'7',3)
        Intern % glob % Intern freq Glob freq  p.value v.test
love      10.277  0.657          26        33   0.000 11.018
want       7.115  0.617          18        31   0.000  8.101
freedom    6.324  0.756          16        38   0.000  6.725
go         4.348  0.378          11        19   0.000  6.236
people     6.324  1.253          16        63   0.000  5.341
music      3.162  0.358           8        18   0.000  4.745
church     2.767  0.318           7        16   0.000  4.384
things     2.767  0.557           7        28   0.001  3.383
health     4.743 12.177          12       612   0.000 -3.996
```

6

Multiple factor analysis for textual data

6.1 Multiple tables in textual data analysis

An important part of survey data analysis is the relationships between answers to different questions. When these questions are open-ended, we can build a documents × words table for each. Documents can be, as we have seen, each individual's free text answer or answers aggregated by category of respondent (see Figure 6.1, where each table corresponds to a question).

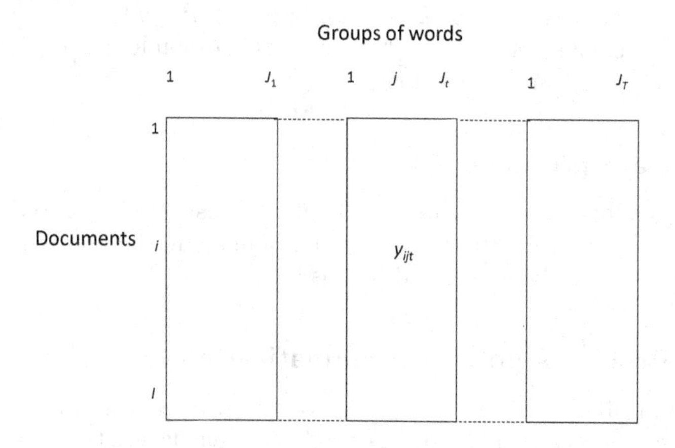

FIGURE 6.1
A set of I documents described in terms of T groups of words. Group t is made up of J_t words.

Other types of studies can lead to similar issues. We may want to compare the evolution over time of a given topic, e.g., the economy section of a range of newspapers. We can do this as follows. For each newspaper, we aggregate the articles according to the qualitative variable *weeks* (or months) and build the weeks × words ALT. Then, these different tables are binded row-wise. A comparison of such lexical tables can then be made as for the previous case. Alternatively, we might want to compare the translations of European Union texts into different languages, or answers to the same question, aggregated according to the same categories but coming from different samples, possibly in different languages. Indeed, this is the type of example we are going to use to introduce the method presented here. Comparing such tables—while also analyzing them all together at once—is the theme of this chapter. We will see that MFACT is a good way to do this. Applied to a multiple table, it can uncover directions of inertia common to all tables, as well as those which—on the contrary—are specific to certain ones. In addition, this method makes it possible to jointly analyze textual and contextual data by giving the two data types symmetric roles in the analysis.

6.2 Data and objectives

To introduce MFACT and its properties, we will again work with the *International Aspiration* survey. We will use responses to the question: *What is most important to you in life?*, and the follow-up: *What are other very important things to you?*, from the UK (*Life_UK*), France (*Life_Fr*), and Italy (*Life_It*). In these three countries, 1043, 1009, and 1048 respondents, respectively, answered equivalent survey questions.

6.2.1 Data preprocessing

A careful spell-check was performed on all the answers. Recall that the word *petitsenfants* (grandchildren) was created manually in the French answers, for these two words to be considered just as one.

6.2.2 Problems posed by lemmatization

For this multilingual study, lemmatization could be useful to erase the influence of specific syntactic features of different languages. Unfortunately, the freely available lemmatization package **TreeTagger** does not perform well on free text answers which are not necessarily grammatically correct. For example, we find answers:

- Without a verb: "*just the memory of my last husband, law and order, welfare for old age people*".
- With only an infinitive: "*d'avoir du travail et pouvoir vivre de son travail*" (to have work and a living wage).
- Composed of a single subordinate clause: "*que mes enfants et mes petitsenfants soient heureux*" (that my children and grandchildren be happy).
- Which are complex but do not follow all grammatical rules: "*job, being a teacher, I love my job, for the wellbeing of the children*".

Consequently, we analyze the actual responses here. Note that the results differ little from those obtained from lemmatized corpora (see Section 6.5). Moreover, working from non-lemmatized corpora can provide nuance when taking into account the meaning of texts. Recall that if two inflected forms of the same lemma are distributional synonyms, they play exactly the same role, whether or not they correspond to separate columns. If, on the other hand, they are not distributional synonyms, it means that they are used in different contexts that give them distinct meanings; regrouping them in a single lemma would mask this information.

The main problem comes from infrequent inflected forms which, when not grouped together, are not taken into account, whereas if they had been grouped in the same lemma, would have been retained. This problem is particularly relevant in the case of a small corpus. It is however possible to lemmatize a corpus before analyzing it with the **Xplortext** package; this is the case for the corpus analyzed in Section 6.5.

6.2.3 Description of the corpora data

The three corpora (*Life_UK*, *Life_Fr*, and *Life_It*) have respective sizes 13,917 (1334 distinct words), 14,158 (1243 distinct words), and 6156 (785 distinct words).

6.2.4 Indexes of the most frequent words

Initial information can be gleaned from reading through the word indexes. These show the use of numerous homologous words (in terms of meaning) from one language to the next. In this way, we find the words *health*, *work*, and *family* among the most-used in the three languages.

Nevertheless, the frequencies of such homologous words can vary greatly; for instance, *family* is used 705 times by the British, *famille* 321 times by the French, and *famiglia* 439 times by the Italians (these base frequencies are comparable here because the number of respondents is similar from one sample to another). Note also that *amore* is used 178 times by the Italians, *amour* 88 times by the French, and *love* only 33 times by the British (including the noun and the corresponding infinitive of the verb). On the other hand, it is the British who mention most often their husband (96 times) or wife (76 times), while in French, *mari* is used only 44 times and *femme* 14 times, and even less so by the Italians (*marito* 6 times, *moglie* 5 times). The importance given to love and to spouses does not therefore seem to go hand in hand (in this survey at least). It should be noted here that MFACT, just like CA, will not take into account differences in frequency but rather, possible differences in distribution across documents.

Over and above these initial examples, the high number of homologous words suggests that there are numerous concerns shared by the three samples, which is a necessary condition for a comparative study. Of course, it will also be interesting to discover whether certain themes are country specific, or at least mentioned to a greater or lesser extent in certain countries.

6.2.5 Notation

For each of the T samples (= 3 here), we construct an ALT $\mathbf{Y}_t$ $(t = 1, \ldots, T)$ (see Figure 1.1, Chapter 1), grouping together responses according to the six categories of the `Sexe_Age` variable. When building these tables, tool words are removed using the French and Italian stoplists provided by the **tm** package. As for the English stoplist from the **tm** package, it is modified by removing the infinitive verbs *to have*, *to be*, *to do*, and *to be able*, as well as the gerunds *having* and *being* (though the form *to* remains in the stoplist). The infinitives and gerunds of the French and Italian versions of these verbs are not in the respective stoplists.

In the UK and French cases, words used in at least 10 different responses are retained. In the case of Italy, this frequency threshold is lowered to 5. This leads to keeping, respectively, 137, 116, and 111 different words from the UK, French and Italian corpora.

The table $\mathbf{Y}_t$ has as many rows as there are documents (I rows; in our example I =6), and as many columns as there are words retained from corpus t (J_t; in our example $J_1 = 137$, $J_2 = 116$, and $J_3 = 111$).

In row i and column (j, t), we have the frequency y_{ijt} with which document i (the set of responses corresponding to category i of `Gender_Age`) uses the word j in country t. A multiple contingency table $\mathbf{Y}$ of size $I \times J$, where $J = \sum_t J_t$, can then be built by binding row-wise the tables $\mathbf{Y}_t$. In order for an overall analysis of this multiple table to have any meaning, the documents in each table corresponding to the same row need to be homologous. This is the case here, where categories of the variable `Gender_Age` are the same for each country.

We then turn $\mathbf{Y}$ into a table of relative frequencies, denoted $\mathbf{F}$, by dividing each entry by N, the total number of entries in $\mathbf{Y}$. Thus $\mathbf{F}$, which is the same size as $\mathbf{Y}$, has entries of the form:

$$f_{ijt} = \frac{y_{ijt}}{N},$$

where f_{ijt} is the proportion of the time that document i, $\{i = 1, \ldots, I\}$ is associated with word j, $\{j = 1, \ldots, J_t\}$ in table t, $\{t = 1, \ldots, T\}$. We thus have that $\sum_{ijt} f_{ijt} = 1$. The margins of $\mathbf{F}$ and its subtables, noted $\mathbf{F}_t$ (see Figure 6.2), are defined by replacing the index or indices over which summation is performed by a dot. The general term for the column margin of $\mathbf{F}$ is $f_{i..} = \sum_{jt} f_{ijt}$, and for the row margin, $f_{.jt} = \sum_i f_{ijt}$. The sum of all terms of each $\mathbf{F}_t$ is given by $f_{..t} = \sum_{ij} f_{ijt}$.

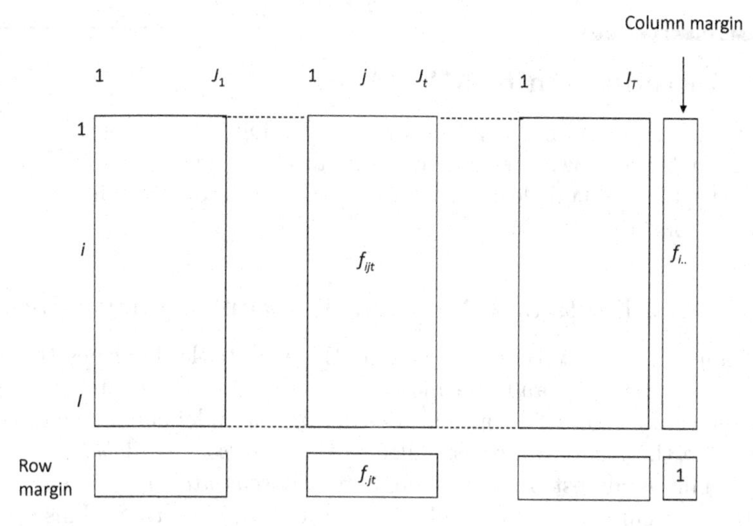

FIGURE 6.2
Multiple contingency table in the form of a table of relative frequencies concatenating the three aggregate lexical tables. Row and column margins are also shown.

6.2.6 Objectives

We would like to analyze this multiple contingency table in order to describe the document categories (of `Gender_Age`) from an overall point of view, i.e., from the point of view of all the tables together, but also from the partial point of view of each table (= country). In what follows, we use the word *categories* in the *document categories* sense, i.e., documents made up of all the answers from a given category. We want to identify inertia axes shared by all tables, some tables, or simply found in one table only. Here are some of the questions we want to answer:

- Which words characterize a given category?
- Which categories characterize a given word?
- Which words are similar in terms of usage by different categories?
- Which categories are similar in terms of vocabulary:
  - for the whole set of tables (shared structure)?
  - in only one table or a subset of tables (specific structure)?

6.3 Introduction to MFACT

MFACT is a method for simultaneously analyzing a set of contingency tables with homologous rows, presented in the form of a multiple table. Before introducing the method, let us first recall the limits of CA when it comes to multiple contingency tables.

6.3.1 The limits of CA on multiple contingency tables

In Chapter 2, we saw that CA on a contingency table describes the gap between observed data and the independence model. Performing CA on the multiple table **F** therefore means considering the independence model calculated from the *overall margins*, called in this context the *global independence model*. This is almost always different to the *separate independence models*, i.e., those calculated from each of the separate tables' margins. This is because of differences between the mean column profiles of the separate tables, due to different relative lengths of the partial documents, i.e., those defined by the rows limited to—successively—each table. In the case of survey data, this is mainly because the proportion of respondents in each category differs from one country to another, most often because of small differences in sampling in each country.

Thus, CA on a multiple contingency table analyzes deviation in column profiles from the global mean column profile, not from the mean column profile of the table to which these columns belong. This CA decomposes the *total inertia* of the multiple table. However, this total inertia is the sum of the *within-table inertia*, due to the differences between the column profiles and the mean column profile of the corresponding table, and the *between-table inertia*, due to the differences between the mean column profiles of each of the tables and the overall mean column profile. Structure that is uncovered depends on both components of the inertia, without us being able to know the contribution of each, though the between-table inertia does not typically involve any particular structural features.

In addition, the overall structure of the rows, as manifested in the CA of the multiple table, may be influenced more by one of the tables than the others. Indeed, the relative importance of each of the tables in the analysis, measured by the sum of the contributions of its columns, is not the same. In the case of contingency tables, the predominant role of table t can have two origins:

- Strong structure in table t, i.e., a strong relationship between rows and columns, which leads to high eigenvalues; all things being equal, the stronger its structure, the more a table influences the overall analysis.
- A high global proportion ($f_{..t}$), which could be due, e.g., to a large sample size, which would lead to large weights for the columns of table t.

This information, however interesting in itself, must not hide the differences between lexical profiles within each of the tables, which is what interests us most in this setting. We also note that the CA of the multiple table does not provide information on two important points:

- The comparison of the distances between documents according to the tables; thus, we cannot know which documents are close to each other from one table's point of view, and far apart from another's.
- The similarity between the groups of columns (a group=a country), considered as a whole, in the sense of the similarity between the document configurations which are induced by their column words.

6.3.2 How MFACT works

Here, we briefly present MFACT in the textual data science setting. This method is an extension of CA which makes it possible to analyze multiple contingency tables, i.e., tables whose row documents are described in terms of several groups of column words, whilst retaining—as much as possible—the point of view and properties of CA. MFACT combines *internal CA* (ICA), a method proposed independently by Pierre Cazes, on the one hand, and Brigitte Escofier and Dominique Drouet, on the other, and multiple factor analysis (MFA), proposed by Brigitte Escofier and Jérôme Pagès. A consequence of adopting the principles of ICA is that a word's profile is not compared with the mean word profile of the overall table, but with the mean word profile of the table it belongs to. As for the MFA approach, it helps balance the influence of each of the tables when looking for the first global axis, and provides tools to compare the different tables.

Thus, MFACT provides us with results:

- Analogous to those from CA—principally how words are represented and the global representation of documents, induced by the whole set of words, which can be interpreted like those from CA.
- Specific to multiple tables, in particular the superimposed representation of the global and partial configurations of the documents. The partial configurations are those induced by each group of words.

Furthermore, MFACT inherits interpretation aids from MFA, whose form we will describe with the help of the *International Aspiration* example.

6.3.3 Integrating contextual variables

We can also introduce quantitative or qualitative contextual variables into the analysis by giving them an active or illustrative role. Here too an MFA approach can help balance the influence of each table.

6.4 Analysis of multilingual free text answers

We apply MFACT to the multiple table constructed in Section 6.2.5. Thus, three active groups are taken into account, corresponding to the words from each country.

6.4.1 MFACT: eigenvalues of the global analysis

The first global eigenvalue can take values between 1 (= all axes obtained in the separate analyses are orthogonal to each other) and T, the number of groups, (the case where the first axes of the separate analyses are the same, and identical to the first global axis calculated using MFACT). In our example, the first eigenvalue is 2.91, close to the maximum value (the number of groups), i.e., 3. This means that the three first axes of the three separate analyses of the three tables are strongly correlated with each other and with the first axis of the global analysis. This axis is a dominant direction of inertia with 34.3% of the total inertia, and shared by the three countries (see Table 6.1).

The contribution of a country to the inertia of each axis of the global analysis is the sum of the contributions of the column words of this country, which comes as a percentage. In our example, the columns of the UK, France, and Italy tables contribute, respectively, 34.04%, 32.43%, and 33.54% of the total inertia (= 2.91) (see the results on the group contributions in Table 6.1). The contribution of the three tables to the inertia of the first global axis is therefore balanced.

TABLE 6.1
First five eigenvalues and group contributions to the inertia of the first five axes.

```
> round(res.mfact$eig,3)
       eigenvalue percentage of variance cumulative percentage of variance
comp 1      2.908                 34.252                            34.252
comp 2      1.971                 23.219                            57.471
comp 3      1.463                 17.228                            74.699
comp 4      1.155                 13.605                            88.304
comp 5      0.993                 11.696                           100.000

> round(res.mfact$group$contrib,3)
    Dim.1  Dim.2  Dim.3  Dim.4  Dim.5
UK 34.036 39.406 32.408 29.944 40.125
Fr 32.427 39.293 29.580 36.564 26.906
It 33.537 21.302 38.011 33.492 32.968
```

6.4.2 Representation of documents and words

MFACT builds optimal spaces for representing either documents or words. MFACT global plots can be interpreted in the same way as CA ones: the coordinate values of a document (here, a document is the answers from a Sexe_Age category) or a word are its values for the global factors. As in CA, it is legitimate to superimpose the word and document configurations. An interpretation of the relationship between them rests on the transition formulas which, like in CA, link the coordinate values of documents and words.

The transition formula that connects, on axis s, the position of document i to the positions of all the words (j, t) is similar to that of classical CA: modulo a constant, each document is located at the CoG of the words it uses:

$$F_s(i) = \frac{1}{\sqrt{\lambda_s}} \sum_{t=1}^{T} \frac{1}{\lambda_1^t} \frac{f_{i.t}}{f_{i..}} \left(\sum_{j=1}^{J_t} \frac{f_{ijt}}{f_{i.t}} G_s(j,t) \right).$$

In other words, a document is attracted by the words that it uses and repelled by those it does not. The more documents use words similarly across all tables, the closer they will be to each other.

The transition formula which links, on axis s, the position of the (j, t)th word to those of all the documents, differs slightly to the CA one:

$$G_s(j,t) = \frac{1}{\sqrt{\lambda_s}} \left(\sum_{i=1}^{I} \left(\frac{f_{ijt}}{f_{.jt}} - \frac{f_{i.t}}{f_{..t}} \right) F_s(i) \right).$$

As the coefficient $(f_{ijt}/f_{.jt} - f_{i.t}/f_{..t})$ of $F_s(i)$ can be negative, words are not found at the CoG of documents, except when the document weights are the same for all sub-tables (i.e., $f_{i.t}/f_{..t} = f_{i..}$ for all countries).

This coefficient measures all of the deviations of the profiles of the column words from the column margin profile of the corresponding table. A word is attracted (resp. repelled) by the documents that use it more (resp. less) than if there were independence between the rows and columns of the table.

The proximity between two words from the same table increases the more the two profiles are similar, exactly as in classical CA. The proximity between two words from different tables increases the more they move away—in a similar fashion—from the average profile of the tables to which they belong.

Remember that on the plots, the column margins of each table are at the global CoG, and we only see the deviation from the average column profile of the corresponding table. Thus, the distance between words from different tables is interpretable.

For example, the words *work*, *job*, *travail*, and *lavoro* are found close to each other in Figure 6.4, which means that their profiles diverge in a similar way from the mean column profiles of their groups. These words are in fact *within-language* distributional synonyms (in the case of *work* and *job*) or *between-language* ones in the case of the other pairs of words.

The global representation of the documents provided by MFACT shows that the first factorial axis corresponds to `Age`, and the second to `Gender`. The $>$`55 years` old points, both for men and women, are far from the `31-55 years` old ones, much more so than the latter are from the $\leq$`30 years` old ones. This means that around the age of 55, vocabulary use changes in a marked way, especially for men.

The position of a given document category depends, via the transition formulas, on the whole set of words, both those it attracts and those it repels (or uses much less than the average). Similarly, the position of a word depends on the relative importance of each document in its profile.

In addition to these words, many other words homologous from one language to another are close to each other on the plot. This proximity expresses similar distributions of concerns by age group in the different countries. Just like in CA, the absolute frequency of words, which varies from one country to another, does not influence their proximity; only the profiles are taken into account.

Of course, certain homologous words may be far from each other or not present in one of the samples. For instance, *daughter*, *fille*, and *figlia* (see Figure 6.4) lie at entirely different positions on the principal factorial plane. These words are used, respectively, 10, 2, and 0 times by the men, and 11, 9, and 5 times by the women. These actual frequencies are in fact comparable, as the number of respondents from each gender is of the same order of magnitude. The word *daughter* is a more *masculine* one (in the sense that it is principally used by men) than its French and Italian homologs.

A lexical explanation of the respective positions of the categories comes from looking closer at the words plot. As showing all 364 words would be illegible, it is often useful to draw several plots. In this example, we have chosen to plot the highest-contributing words (see Figure 6.3—bottom), those dealing with family (see Figure 6.4—top), and certain words to do with general worries, such as *travail* (work), *bonheur* (happiness), and *entente* (harmony), (see Figure 6.4—bottom). In general, useful plots in a given analysis will depend on the questions being asked.

Interpretation of these plots is based on, as in CA, the transition formulas. Thus, to the right of the first axis, we find the words over-used by the young, and on the left, those favored by older people. Words in the negative part of the second axis are *feminine*, whereas those in the positive part are *masculine*. Furthermore, vocabulary evolves with age. For instance, the young talk of work (*work*, *job*, *travail*)—especially the men, and family (*family*, *famille*, *famiglia*, *daughter*, *son*, etc.)—especially the women. The oldest respondents talk more of their grandchildren (*grandchildren*, *petitsenfants*) and harmony (*entente*, *accordo*) between family members. The youngest aspire to happiness (*happiness*, *happy*, *bonheur*, *heureux*, *felicità*), the oldest to peace of mind and tranquility (*peaceofmind*, *tranquillité*, *tranquillità*, *paix*, *pace*). Tranquility is a particularly *masculine* feature in the French sample.

The positions of the words *wife*/*femme*/*moglie*, on the one hand, and

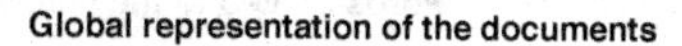

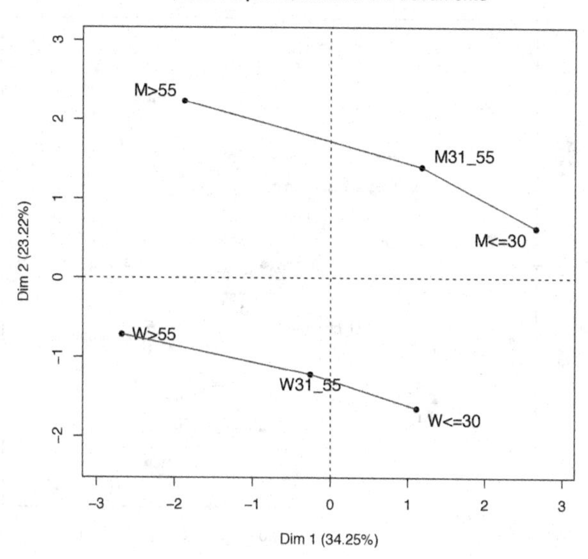

Representation of the meta-keys

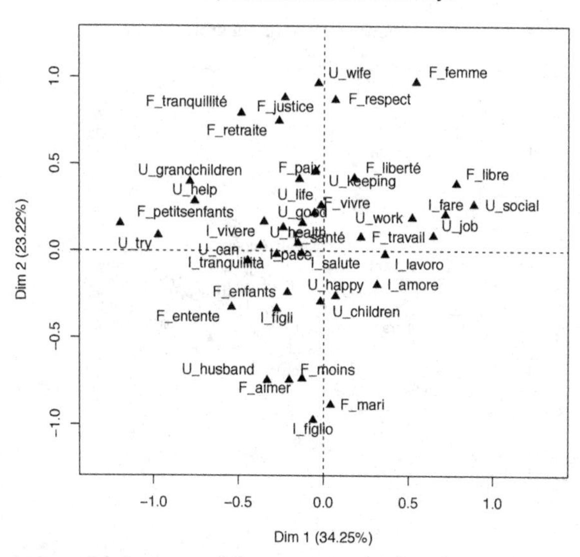

FIGURE 6.3
Plane (1,2) of the MFACT. Top: trajectories of men and women in terms of age categories. Bottom: words with a contribution over twice the average. F, U, and I identify which country words belong to.

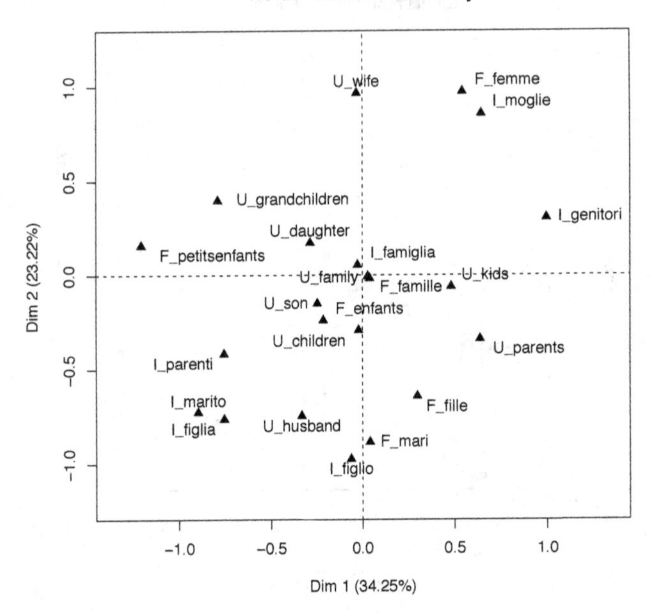

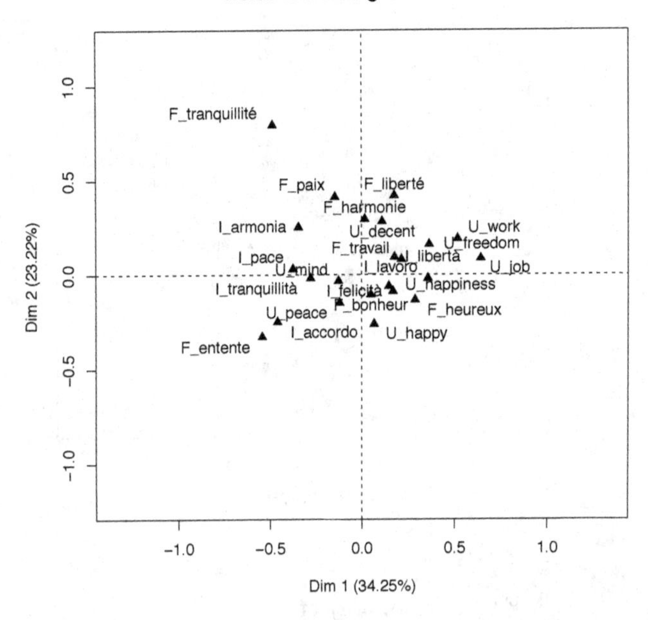

FIGURE 6.4
Plane (1,2) of the MFACT. Top: words related to the family. Bottom: words related to general worries.

husband/mari/marito, on the other, provide interesting information. For instance, the use of words homologous to *wife*, an almost entirely *masculine* word:

- Is not very associated with using other words related to the family.
- Corresponds to the two youngest male age groups in the French and Italian tables, but is especially used by the two oldest age groups in the UK table (see Table 6.2).

In addition to this, the use of words homologous to *husband*, an almost entirely *feminine* word:

- Is highly associated with the use of other words related to the family.
- Does not correspond to the same age groups from one table to another; it is associated with the two youngest age groups in the French table, essentially shared by all of the age groups in the UK table, and only used by the two oldest age groups in the Italian table (see Table 6.2).

TABLE 6.2
Profiles of words homologous to *husband* and *wife* in the three languages.

	husband	wife	mari	femme	marito	moglie
H<=30	0.00	9.21	2.27	35.71	0	40
H31_55	3.12	39.47	0.00	42.86	0	40
H>55	1.04	39.47	0.00	21.43	0	20
F<=30	21.88	1.32	36.36	0.00	0	0
F31_55	41.67	3.95	56.82	0.00	50	0
F>55	32.29	6.58	4.55	0.00	50	0
Sum	100.00	100.00	100.00	100.00	100	100

Numerous other remarks could be made about word use by each category. One example is the large distance separating the Italian words *genitori* (parents) and *parenti* (relatives)—see Figure 6.4. Clearly these are far from being synonyms or equivalent in this corpus.

The (1,3) plane (not shown here) exhibits the Guttman effect. Note that this effect can also be seen on the (1,3) plane of the UK and French samples analysed individually, and the (1,2) plane of the Italian sample.

6.4.3 Superimposed representation of the global and partial configurations

The global (or average) configuration of the categories (see Figure 6.3) is constructed using all groups of words. Partial configurations of categories can also be obtained; these are constructed one by one, separately, using their specific groups of words, then superimposed on the planes of the global analysis.

These different configurations can be shown on the same plot, which makes it possible to compare them.

In this plot, the partial configurations are dilated by a coefficient T equal to the number of tables. Thus, each global document, whose coordinate values are calculated from all the words it uses, is located at the CoG of the corresponding partial documents, whose coordinate values are calculated from the words of each country alone. In other words, each global point can be seen as an *average* point.

In our example, the superimposed representation of the partial and global document configurations shows a very strong resemblance between the three partial ones. For each country, the first axis contrasts the youngest with the oldest, while the second contrasts men and women. This is thus another way of highlighting shared features.

Nevertheless, beyond the overall resemblance of the three partial configurations, we do see some differences. As is visible in Figure 6.5, the subcloud formed by the partial documents corresponding to `men>55` has a high within-cloud inertia, in comparison with partial document subclouds corresponding to the other categories. In particular, the `men>55` from the French and UK tables lie in positions particularly far from the CoG, unlike the Italian's.

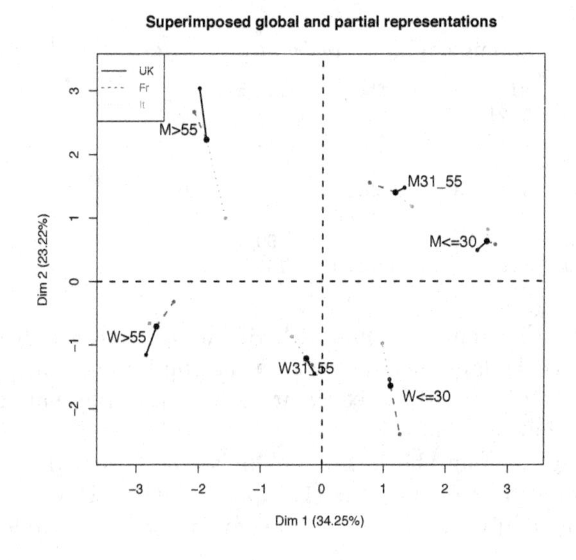

FIGURE 6.5
Superimposed global and partial representations of the document configurations.

As the differences between the partial documents occur mainly along the second axis, this suggests that the vocabulary of men in this age group differs markedly from that of women of the same age in these two countries.

In contrast, for the Italian table, these two categories involve fairly similar vocabulary. It is possible to relate this to the presence of French and English words in only the upper part of the second quadrant.

Similarity between the three partial clouds can be evaluated globally in terms of the ratio of inertia (see Table 6.3), calculated as follows. The partial points cloud i^t, for $\{t = 1, \ldots, T\}$ and $\{i = 1, \ldots, I\}$, is divided into I clusters, where the T partial points related to the same document belong to the same cluster (here, T=3). The ratio of between-class inertia to total inertia indicates to what extent structure seen in axis s corresponds to structure shared by all groups. If axis s reflects a shared structure, the partial points related to the same document will be close to each other, the within-class inertia low, and the between-class inertia close to the total inertia (and thus the ratio of inertia will be close to 1). Otherwise, the ratio will be small. For a given t, the distances between the partial points related to the same document i^t and the mean point i on axis s indicate the contribution of this document to the within-class inertia. We are particularly interested in:

- Row documents whose corresponding partial points are close to each other (very low within-class inertia); such documents illustrate the shared structure the given axis represents.
- Row documents whose corresponding partial points are very spread out (high within-class inertia); these are exceptions to the common structure, and studying them can provide interesting information.

TABLE 6.3
The ratio of between-inertia to total inertia, axis by axis.

```
> round(res.mfact$inertia.ratio,3)
Dim.1 Dim.2 Dim.3 Dim.4 Dim.5
0.987 0.906 0.943 0.917 0.943
```

We can also globally compare the three configurations corresponding to the three partial clouds. The correlation coefficients, calculated for each axis between the documents' *global* and *partial* factors, help us to refine this similarity measure between partial configurations (see Table 6.4). These coefficients are called *canonical correlation coefficients*. If, for a given axis, all of these coefficients are high, then it corresponds to a direction of inertia common to all of the groups. Otherwise, it represents a direction of inertia present only in groups for which there is high correlation.

In our example, the first five MFACT inertia axes can be seen as shared by all three groups of columns (= the three countries).

TABLE 6.4
Canonical Correlation coefficients. First five axes.

```
> round(res.mfact$group$correlation,3)
   Dim.1 Dim.2 Dim.3 Dim.4 Dim.5
UK 0.997 0.987 0.984 0.951 0.986
Fr 0.990 0.986 0.952 0.954 0.989
It 0.993 0.958 0.986 0.977 0.978
```

6.4.4 Links between the axes of the global analysis and the separate analyses

The first step in MFACT involves *separate analyses* of each of the tables. This step, critical for calculating the column weights in the overall analysis, is automatically executed by the `MFA` function from the **Xplortext** package. The factors on the documents in these separate analyses (called *partial axes* in this context) can be projected as supplementary variables on MFACT factorial planes. They are thus represented in terms of their correlation with the global factors of the MFACT (see Figure 6.6).

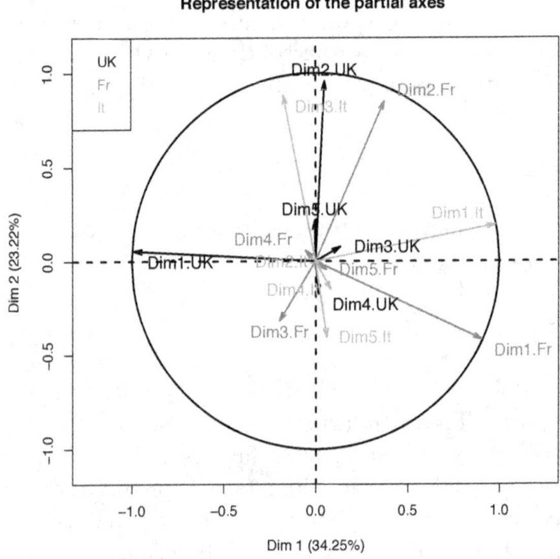

FIGURE 6.6
Representing the factors of the per-table analyses. DimS.UK is the rank *s* factor of the UK group's analysis.

In this example, the first MFACT factor is strongly correlated (either positively or negatively, it does not matter which) with the first factor of each of the separate analyses. This highlights the fact that the first global axis corresponds to a *mean* or *compromise* between the first axes of the separate analyses. As for the second global axis, it is highly correlated with the second axes of the separate analyses of the UK and French tables, and the third axis of the Italian table.

Hence, the first MFACT plane shows fairly similar structure over the documents to that displayed on the first plane of the separate analyses of the UK and French tables, and the (1,3) plane of the Italian one.

6.4.5 Representation of the groups of words

We may also be interested in a global representation of the groups of columns that can answer the question: "Do groups of words define similar configurations over the documents?" In other words: "Are the distances between documents similar from one table to another?" MFACT offers a representation of the groups in which each of them is shown as a single point. On this plot, two groups of words (= two countries) are close if each induces a similar structure over the documents. Here:

- Axes can be interpreted as they are in the global analysis.
- The coordinate value of a group on an axis is equal to the sum of contributions of words in the group to the axis' inertia (= group contribution); this is always between 0 and 1.
- A small distance between two groups on axis s means that there exists in these groups a direction of inertia of a similar magnitude in the direction of the global axis s, low inertia if their coordinate values on axis s are small, and high inertia if they are large. Thus, this representation shows the similarity or dissimilarity of groups from the point of view of the global axes.

In Figure 6.7, a group of variables is represented by a single point. In our example, the three active groups have high coordinate values—close to 1—on the first factorial axis. These values are equal to the contribution of each group to the inertia of the first global axis. Clearly, this axis corresponds to a direction of high inertia in the three word clouds. We can therefore say that differences in vocabulary are strongly age-related in all three countries. The second axis (which we have interpreted as an axis contrasting the genders) is a direction of high inertia in the case of France and the UK, very likely because of the greater difference between the older male and female categories in these countries, as shown in the superimposed representation of the global and partial clouds (see Figure 6.5).

6.4.6 Implementation with Xplortext

The code in `script.chap6.a.R` does the following:

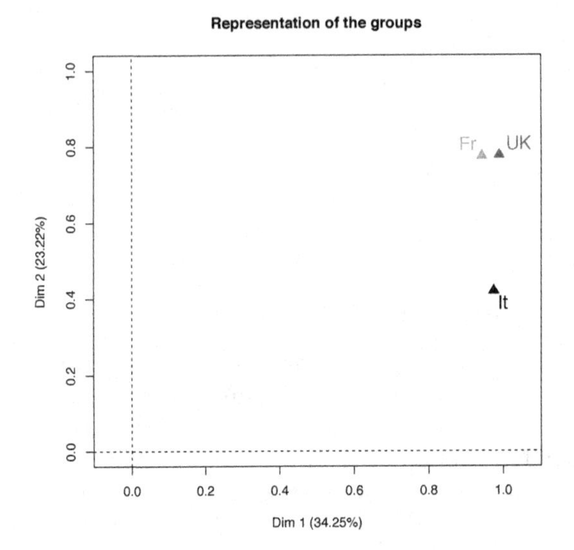

FIGURE 6.7
Representation of the groups of column words.

- The `TextData` function is run on each of the three databases in order to construct the three ALT with respect to the variable `Gender_Age`.
- The three ALT are then binded row-wise.
- The `MFA` function of the **FactoMineR** package then performs MFACT on the resulting table.
- The `plot.MFA` function is then used to obtain the desired plots.
- In addition, certain results are output and the profiles of the 6 words homologous to men and women, respectively, are calculated and displayed.

6.5 Simultaneous analysis of two open-ended questions: impact of lemmatization

To study the impact of lemmatization, it would seem appropriate to look at an example in a language whose verbs and adjectives have many inflected forms. In such languages, it is likely that the impact will be greater than in, for instance, English.

Here, we look at an example in French. It comes from a post-election survey in Belgium. The survey had two open-ended questions: "What does the left mean to you?" and "What does the right mean to you?" along with numerous other typical questions for a post-election survey. We focus on the results from the Wallonia region involving 1423 respondents.

6.5.1 Objectives

The main aim here is to compare the structure defined over the respondents by, respectively, the lemmas and graphical forms corresponding to either what the left or what the right means to them. Another goal is to compare the structure associated with each of the two questions.

6.5.2 Preliminary steps

First, a careful spell-check and correction of free text answers is performed. This is important if we want to lemmatize the responses, as we do here. The subsequent lemmatization was performed by Dominique Labbé with the help of his **Lexicométrie** software package.

This package also *tags* each lemma with one of Nf=feminine noun, Nm=masculine noun, Ver=verb, Adj=adjective, or Adv=adverb. We thus end up with the following corpora: two constructed with respect to the free text answers about the left, seen as either sequences of graphical forms (*left-forms*) or sequences of lemmas (*left-lemmas*), and correspondingly, two corpora of respectively *right-forms* and *right-lemmas* for the right. Note that for both questions, lemmatization leads to a smaller number of distinct words (see Table 6.5).

TABLE 6.5
Summary data on the four corpora formed from the two groups of free text answers, given either in graphical form or as sequences of lemmas.

	Left graph. forms	Left Lemmas	Right graph. forms	Right Lemmas
Occurrences	11554	11940	10090	10310
Distinct words	1508	1224	1465	1215
Retained words	86	94	98	102

Note that tool words have been removed, both in the lemmatized and non-lemmatized corpora. Words used in at least 10 responses are retained. Furthermore, the lemmas *savoir* (to know), *rien* (nothing), *aucun* (none of), *idée* (idea), and *non* (no), as well as the corresponding graphical forms, were removed. These words were in the expressions *je ne sais pas* (I don't know), *non, je ne sais pas* (no, I don't know), *il ne sait pas* (he doesn't know), *rien* (nothing), and *aucune idée* (no idea). Thus, respondents who use only one of the expressions listed above are considered non-respondents, since their responses become empty, which means that some information is lost. However, the words listed above are used copiously, and if we keep them, they alone determine the first factorial axes and hide the information in the answers which corresponds to definitions of the left and right, which interest us more.

After all of this has been done, only 939 respondents have non-empty responses overall, and 484 have empty responses. This suggests that expressing an opinion on this subject is difficult, even intimidating. It would be interesting, but beyond the scope here, to study the characteristics of non-respondents.

6.5.3 MFACT on the left and right: lemmatized or non-lemmatized

Next, we build four LTs, all of which have 939 row documents, and respectively 86 graphical forms relative to the left (first table), converted into 94 lemmas (second table), 98 forms relative to the right (third table), converted into 102 lemmas (fourth table) (see Table 6.5). In comparison to the "right" responses, the "left" ones involve more distinct words and less retained words, both for the graphical forms and lemmas. We have not looked for what the underlying reasons might be for this. However, note that the "right" answer comes after the "left" one in the survey, and often takes it as a point of reference. This likely leads to lexicological differences that are hard to characterize.

The multiple table to be analyzed binds row-wise these four lexical tables, along with two columns corresponding to the variables `Education` and `Gender_Age`. The two lemma groups, as well as those corresponding to the contextual variables—of course—are considered supplementary groups. The plot of the groups (Figure 6.8) shows that, on the one hand, the left—from both form and lemma viewpoints, and on the other, the right—from both form and lemma viewpoints, have common factors corresponding to the (1,2) axes, with similar inertia (this is also the case on subsequent axes).

We can also see, on the same plot, that left and right have similar structure too. Checking the data confirms that respondents tend to construct both responses (left and right) in the same way (see Table 6.6).

Lastly, the plot shows that the contextual variables considered are not very associated with the vocabulary used by the respondents.

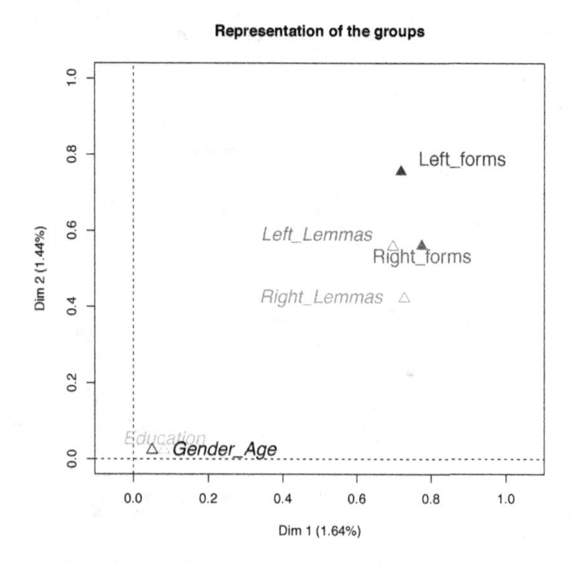

FIGURE 6.8
A graphical representation of the groups. Both groups of graphical forms are active. The two groups of lemmas, as well as those corresponding to the contextual variables, are supplementary.

TABLE 6.6
Examples of responses from three respondents defining the "left" and then the "right".

The left	The right
– la classe ouvrière (working class)	– la classe capitaliste (capitalist class)
– protection des ouvriers (worker protection)	– protection de ceux qui ont des biens (owner protection)
– c'est plus axé sur le social . . . ce sont tous les problèmes sociaux de société (more focused on social aspects . . . all of the social and societal problems)	– c'est plus axé vers le capital, les entreprises, la gestion (more focused on capital, businesses, accounting)

Figure 6.9 plots the graphical forms and the lemmas. It shows that the configurations defined by the lemmatized and non-lemmatized groups are similar, and also that they can be interpreted semantically in the same way. Indeed, the axes contrast—both for the graphical forms and lemmas—definitions of the left and right which correspond to similar viewpoints for these questions. In effect, as we simultaneously analyze the responses to both questions, respondents differ in their general view on these questions and, more generally, on politics.

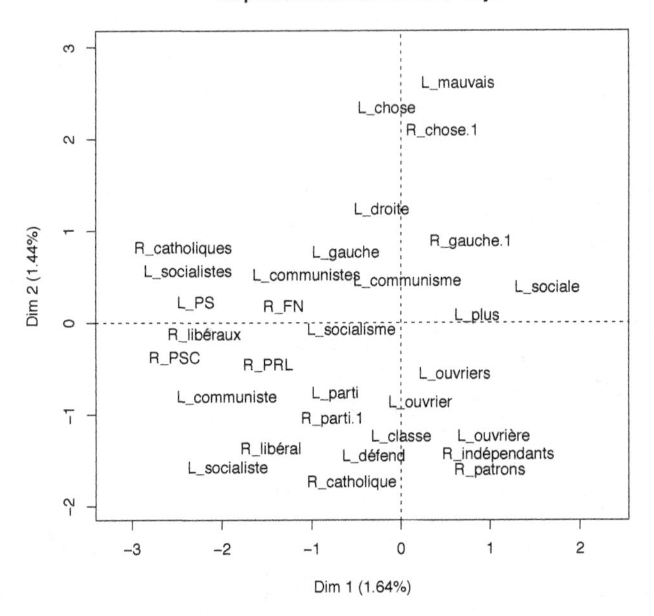

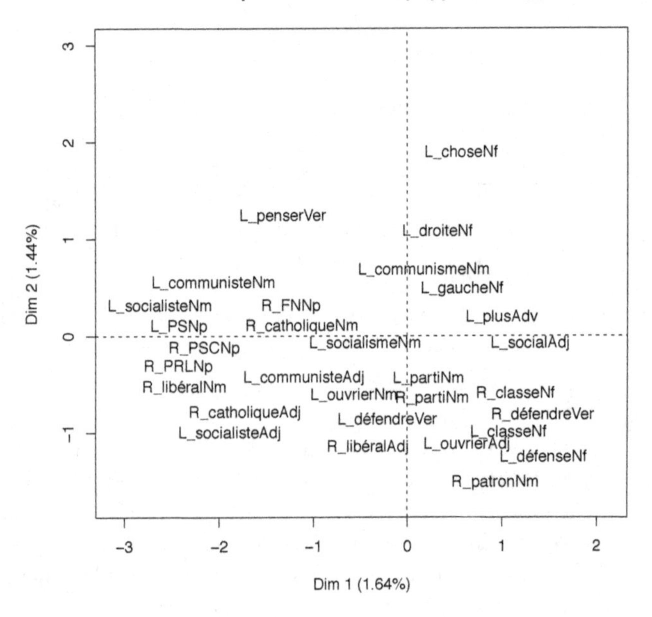

FIGURE 6.9
MFACT (1,2) plane. Top: representation of the meta-keys whose contribution is over twice the mean one. Bottom: lemmas with the best representation quality, projected as supplementary words.

To give an example, the closeness between party acronyms from both the left and right indicates that they are frequently used by the same respondents; consequently, when a respondent refers to the left using acronyms of left-wing parties, they also tend to refer to the right using acronyms of right-wing parties.

As in the case of classical CA, interpretation involves finding the words which contribute most to the axes' inertia. It is not our objective here to interpret the results in detail; we limit ourselves to determining differences across the principal plane.

On the left of the first axis, we find the words *socialistes/catholiques/libéraux* (socialists, Catholics, liberals), and the acronyms PRL, PSC, PS, and FN for, respectively, the Liberal Reformist, Christian Social, Socialist, and National Front parties. The best-represented lemmas, found to the left of the axis, correspond to the lemmatized forms of these words. To the right of the axis, only the adverb *plus* and adjective *social* contribute. This first axis therefore sees the definitions of left and right in terms of overall ideology, as opposed to all other points of view. The second axis contrasts vague words like *chose* (thing), *mauvais* (bad), *gauche* (the left), and *droite* (the right) at the top, with words corresponding to definitions of the left and right in terms of the societal groups they correspond to: *défense* (defend), *défendre ouvriers* (defend workers), *patrons* (bosses), and *classe* (class), at the bottom.

6.5.4 Implementation with Xplortext

The results are obtained using the code provided in `script.chap6.b.R`. We proceed as follows:

- First, two stoplists are created, one corresponding to the stopwords in the `tm` package for the French text turned into lemmas, and the other to words indicating non-responses.

- Four successive calls to the `TextData` function build the four LT which correspond to the non-lemmatized and lemmatized responses to the "left", and the corresponding two for the "right".

- Only row documents which are non-empty for all four lexical tables are retained.

- The multiple table is then built; it binds row-wise the four LT as well as columns corresponding to the two contextual variables.

- The `MFA` function of the **FactoMineR** package then analyzes the multiple table; the groups corresponding to the two lemmatized tables and two contextual variables are declared as supplementary.

- The `plot.MFA` function outputs the plots asked for.

- Specific lines of code are then used to contrast words by group.

6.6 Other applications of MFACT in textual data science

MFACT can be used to address a diverse range of problems in textual data science. Here are a few examples culled from published articles.

- The simultaneous analysis of open-ended and closed questions from the same questionnaire. Indeed, MFACT can be applied to a multiple table composed of subtables of different types: contingency tables, but also quantitative and/or qualitative ones. Tables should be in the usual form for analysis by CA, PCA, or MCA. Necessary coding is carried out by the function `MFA` from the **FactoMineR** package in a manner transparent to users. This method offers a new perspective on the treatment of open-ended questions, by simultaneously having closed and open-ended questions as active variables. A comparison of the partial configurations makes it possible to understand the features that each type of question is capable or incapable of capturing. In addition, this avoids pitfalls encountered in the direct analysis of free text answers alone, in particular the construction of overly unstable axes due to the presence of small groups caused by certain specific responses.
- The joint use of quantitative and verbal information, e.g., as often found in wine guides.
- Highlighting the chronological dimension of a corpus by applying MFACT to the table which binds row-wise the textual table of documents × words with the quantitative vector containing the dates the documents were produced.

6.7 MFACT summary

MFACT helps analyze multiple tables made up of several contingency tables binded row-wise, corresponding to a set of documents (rows) described by several groups of words (columns). Each group of columns is centred on its own margins, and thus this method analyzes divergence from the within-table independence model. Furthermore, MFA approach helps balance the importance of each group in the determination of the first global axis. This method leads to results:

- Analogous to those from CA, i.e., essentially a global representation of row documents and column words.

- Specific to multiple tables, principally the superimposed representation of the configurations induced by each group of words over the set of documents, and the representation of the partial axes, i.e., those obtained in the separate analyses.

We can also bind row-wise contingency tables and quantitative and/or qualitative tables in the same analysis. Here again we follow the principles of MFA to balance the influence of different tables and visualize both the global structure and the partial structure corresponding to each group. The interpretation of results can be assisted by the numerous interpretation aids available for MFA.

7

Applications and analysis workflows

7.1 General rules for presenting results

The standard ways to interpret the results of each method have already been detailed in their corresponding chapters. It is now a question of how to present these results and their interpretation. Suppose we are in a setting where we have used a factorial method and a clustering method together. A standard way to present these is as follows:

1. Introduction: present the study.
2. Textual data and objectives:
    - Constitution of the corpus.
    - Contextual data obtained and why they are of interest.
    - How the data have been encoded and resulting data table(s).
    - Justification of the encoding and perhaps a discussion of other possible encodings that were either put to the side or retained for complementary analyses.
3. Description and justification of the methodology:
    - Sequence of methods used, clearly showing parameter choices made.
    - For the factorial methods: frequency threshold for retaining/discarding words, possible removal of words, justifying decisions made, stating which elements are active and which are supplementary.
    - For clustering documents located in factorial spaces: justification for the chosen method, provision of the criteria used to make the decision on the number of axes to be included in the clustering.
    - Specify the information that will be used to describe clusters: words, and perhaps characteristic documents and/or contextual variables.

4. Results:

    - Basic summary statistics: size of the corpus and vocabulary, average document length, most frequent words, most noticeably absent or infrequent words, and those whose presence is unexpected and thus surprising.
    - Initial analysis of the content from reading the glossaries of words and repeated segments.
    - Factorial analysis (CA or MFACT):
        - –Cramér's V (for CA) and bar plot of the eigenvalues; both for CA and MFACT, the eigenvalues give information about structure in the data and should be commented on,
        - –number of interpretable factorial axes; summarize the information they contain with a short label,
        - –interpret factorial axes, separately or in pairs (i.e., by planes),
        - –plots of the factorial planes; these are very important, and must be sufficiently worked to provide clear information. Case by case, do one or several of the following: only represent points (documents or words) that sufficiently contribute and/or are well represented; display and comment on the trajectories in comparison with what was expected, especially in the case of chronological corpora; if possible, use bootstrap methods to validate the representations; plot supplementary elements also,
        - –go back to the data and comment on concrete details of interest to the intended public.
    - Clustering (hierarchical, by direct partitioning, or hierarchical with contiguity constraints):
        - –number of axes with which clustering is performed, along with a justification of this choice,
        - –if hierarchical clustering is performed (with or without contiguity constraints), indicate the tree cut and justify why it was chosen,
        - –clear yet concise description of each cluster in terms of all of the information known about the documents,
        - –placing the clusters on the factorial planes (usually the principal plane only), either in terms of their CoGs or by showing documents of each cluster using specific colors or symbols,
        - –if the corpus is chronological and we have chosen to do

hierarchical clustering with a contiguity constraint, see if a real temporal trend is visible or not in the results. If there is one, show how the different time periods formed by cutting the tree differ from each other; for this, a pertinent use of characteristic words and hierarchical words could be quite useful. If, on the other hand, there is not, suggest reasons for the lack of a trend.

5. Conclusions.

    - Describe the major results of the study, underlining what was known and not known before commencing; do not forget that finding evidence of proven structure validates the discovery of more novel results.
    - Mention hypotheses that may arise from the analysis and require further analyses and possibly new data.
    - Provide important feedback on the original data.

Clearly, this general plan must be adapted to each new analysis, and also to the style of each practitioner/researcher.

7.2 Analyzing bibliographic databases

This section is the work of Annie Morin.

7.2.1 Introduction

This section illustrates the use of the methods presented in this book on data from bibliographic databases. This type of database is often constructed at the beginning of new research on a subject. A search can be performed on several websites using keywords, but all of the publications uncovered are not necessarily relevant and may be too numerous to all be read. Textual data science applied to a corpus of texts makes it possible to more quickly select relevant subsets of publications. The bibliographic database could also be a research organization or laboratory's internal publications; here the goal would be to analyze associated researchers' publications.

7.2.2 The lupus data

The data that will be examined in this section concerns *lupus*. The abstracts were downloaded from the *Medline* database by Belchin Kostov in the context of a European project aiming to create a website resource for patients suffering from auto-immune diseases. The keywords used to search for abstracts were *lupus* and *clinical trials*. Only articles published between 1994 and 2012 were retained.

Lupus is a chronic autoimmune disease that can affect different parts of the body and different sections of the population. Symptoms as well as severity of the disease vary from person to person. It has no cure, though periods of remission are possible, and is difficult to treat. Research has led to the development of drugs that can improve patient quality of life.

The file ***lupus-506*** consists of 506 publication abstracts from *Medline*. This database is much smaller than those used in real analyses, but the approach remains the same as for databases with thousands or tens of thousands of abstracts. Our goal here is to show how to conduct this type of bibliographic study by means of CA, in line with the work of Michel Kerbaol. The evolution of proposed treatments is a particular focus of this study.

The study will take place over three steps:

1. Initial analysis of the 506 documents in the database (Section 7.2.3).
2. Follow-up chronological analysis of the 19 years in question (Section 7.2.4).
3. Third analysis of the 19 years, retaining as words only those which are drug names (Section 7.2.5).

7.2.2.1 The corpus

As the database is imported from *Medline*, multiple formats are possible. Here, we load it as an Excel file. Each record (row) corresponds to an article. It is described by 6 variables: the title, year of publication, journal name, name of the first author, abstract, and publication year in groups: before 1997, 1998–2000, 2001–2003, 2004–2006, 2007–2009, and 2010–2012.

Here is a partial extract of the file, showing a shortened version of what the first two rows look like:

```
  Title                    Year          Journal            First_Author
1 Academic outcomes... 2012 Arthritis care & research     Zelko F
2 The -2518 A/G poly.. 2012 European cytokine network    Aranda F

 Abstract                          Year_class
OBJECTIVE: To explore academic.....  2010-2012
Systemic lupus erythematosus (SLE).  2010-2012
```

7.2.2.2 Exploratory analysis of the corpus

We have 506 documents with a total of 93,123 occurrences (7079 distinct words). We remove tool words and only keep words that appear at least 15 times and in at least 10 documents. After this, 786 distinct words remain, for a total of 49,049 occurrences in all. At this point, the documents $\times$ words table is of size 506×786.

Table 7.1 provides some initial statistics: the frequency of the 10 most common words (excluding tool words), how many times they appear, and in how many documents.

TABLE 7.1
Index of the ten most frequent words after removing tool words.

Word	Frequency	Number of documents
patients	2273	462
sle	2126	466
disease	864	310
lupus	832	471
activity	592	224
systemic	588	453
erythematosus	542	448
treatment	505	209
study	501	307
group	477	160

The word *lupus* appears 832 times in 471 documents, though *sle*, meaning *systemic lupus erythematosus*—which also refers to lupus, is found 2126 times in 466 documents.

7.2.3 CA of the documents × words table

We now run CA on the documents × words table with 506 rows and 786 columns. The aim is to identify specific research questions.

7.2.3.1 The eigenvalues

The total inertia of either the document or word cloud is 14.75. The first ten eigenvalues, as well as the percentages of explained inertia and cumulative inertia are shown in Table 7.2. The bar plot in Figure 7.1 shows the sequence of decreasing eigenvalues. The first two axes correspond to 3.67% of the inertia, which may seem very low, but is often the case with bibliographic data where the numbers of documents and words retained are high. In addition, the associated eigenvalues of 0.29 and 0.25 indicate that these axes account for sufficiently general situations, for the analysis to be useful. From the third eigenvalue on, further decreases are small, but the corresponding axes could still be of interest to us in the study. Here, in this example we will focus on the first four axes, with the help of meta-keys and doc-keys (described below).

7.2.3.2 Meta-keys and doc-keys

As already mentioned in Chapter 3, Michel Kerbaol defined what are known as *meta-keys* to simplify the study of bibliographic data. For a given axis, meta-keys are made up of words whose contribution to the inertia is above a given

TABLE 7.2
Eigenvalues and percentages of variance.

	Eigenvalue	% of variance	Cumulative % of variance
dim 1	0.29	1.94	1.94
dim 2	0.25	1.73	3.67
dim 3	0.19	1.32	4.99
dim 4	0.18	1.24	6.23
dim 5	0.17	1.14	7.37
dim 6	0.16	1.08	8.45
dim 7	0.16	1.07	9.52
dim 8	0.15	1.02	10.53
dim 9	0.14	0.94	11.48
dim 10	0.13	0.90	12.38

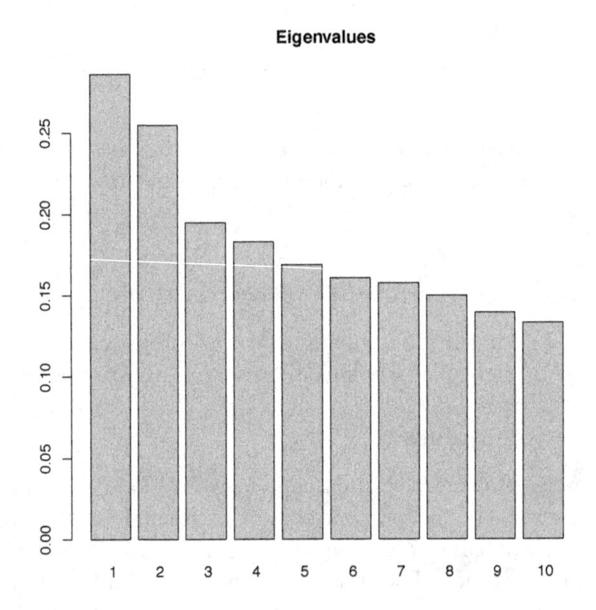

FIGURE 7.1
Bar plot of the first ten eigenvalues.

threshold—more precisely: above the mean contribution multiplied by this threshold. The coordinate values of words found in one of the two possible meta-keys can be positive or negative, and for each axis examined, we will distinguish between both the positive and negative meta-keys. Meta-keys can help us to interpret the CA factors of a lexical table.

In the same way, the documents whose contributions to the inertia of an axis are greater than a certain threshold are found in one of the two possible *doc-keys*. For each axis examined, we will distinguish between the positive and negative doc-keys according to the sign of documents' coordinate values.

By default, in the **Xplortext** package, the threshold mentioned above and used to select meta-keys is set at 3. The threshold is also 3 for the doc-keys. However, here, given the large numbers of words and documents with contributions above 3, we limit the following lists (see Tables 7.3 and 7.4) to words and documents whose contributions are greater than 6 times the mean one.

TABLE 7.3
Meta-keys and doc-keys for the first axis.

Axis	Meta-key	Doc-key	Theme
1 +	*association, allele, gene, polymorphism, susceptibility, polymorphisms, hla, associated, risk, genetic, alleles, controls, sle, genotypes, genes, genotype*	181, 190, 5, 317, 109, 246, 87, 250	genetic engineering
1 -	*placebo, treatment, dhea, months, dose, day, bmd, weeks, therapy*		

The doc-key 1- is empty. If we lower the threshold to 3, the meta-key would be larger and we would have one element in the doc-key corresponding to the abstract 151.

TABLE 7.4
Meta-keys and doc-keys for axes 2, 3, and 4.

Axis	Meta-key	Doc-key	Theme
2 +	*damage, health, disease, sdi, physical, factors, social*	350,441,172,209, 450,290,130,302	quality of life damage index/sdi
2 -	*cells, cell, expression, depletion rituximab, lymphocytes, anti, gene*	280,310,192,140	biotherapies
3 +	*anti, antibodies, dsdna, syndrome, acl*	113,484,410	subgroup studies
3 -	*calcium, bmd, lumbar, group, premenopausal, bone, dhea, allele polymorphism, cholesterol mineral, alone, control, groups, controls, taking, gene, susceptibility*	162,288,343,126, 261,262,181, 8,151,401	bone mineral density
4 +	*calcium, bmd, pulmonary, vascular, mortality, risk, alone, lumbar, death, events, cases*	162,487,253	cohort monitoring
4 -	*cells, health, activity, physical, cell, scores, bilag, social, damage, depletion*	350,290,130	quality of life bilag index scores

The words that strongly contribute to the inertia of the first axis and are located on the positive side are related to genetic engineering: *gene*, *allele*, *genotype*, both in singular and plural forms. A closer look at the doc-key on the positive side of the first factorial axis confirms this. To the negative side, the meta-key contains words related to lupus treatments and protocols.

The meta-key for the positive part of the second axis contains the words *damage, health, disease, physical, sdi, social*, and *factors*. The associated doc-key deals with the socio-economic situation of patients and their quality of life as measured by the *damage index* defined by the Systemic Lupus International Collaborating Clinics/American College of Rheumatology.

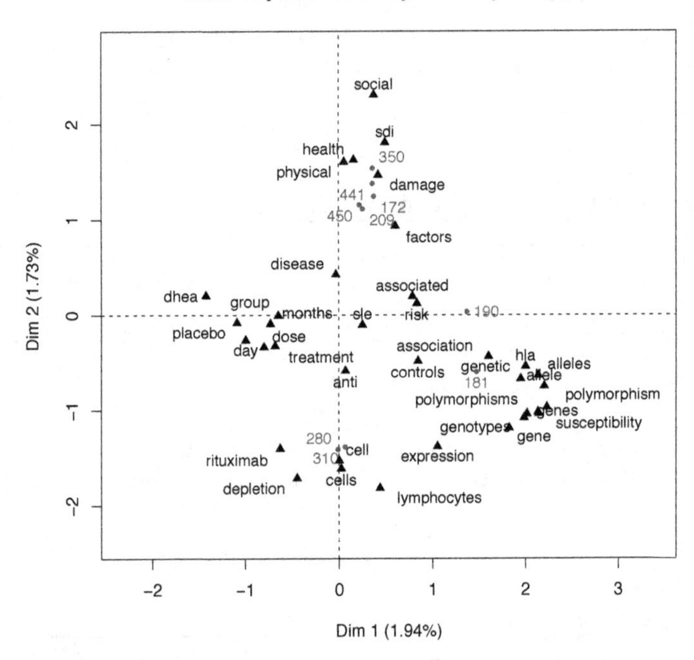

FIGURE 7.2
Meta-keys and doc-keys on the (1,2) factorial plane.

In contrast the 2- meta-key deals with words like *B cells* and *T cells* and associated biotherapies.

Figure 7.2 shows the projections of meta-keys and doc-keys onto the first factorial plane. To avoid overlapping elements on the plot, the threshold for selecting words and documents is set at seven times the mean contribution. This threshold may seem high, but after examining the word lists of the meta-keys, it is clear that the factorial planes will be easier to read and interpret with fewer points. Lowering the threshold increases the number of projected elements.

Figure 7.3 shows the projections of the meta-keys and doc-keys on the (3,4) factorial plane. Axis 3 contrasts, on the negative side, articles dealing with bone mineral density with, on the positive side, articles on studies performed on different types of patients. The 4+ meta-key and the associated doc-key involve publications on cohorts of patients that have followed different protocols, while the 4- meta-key involves the quality of life of patients in relation to *bilag index scores*, which measure the active level of the disease for particular populations.

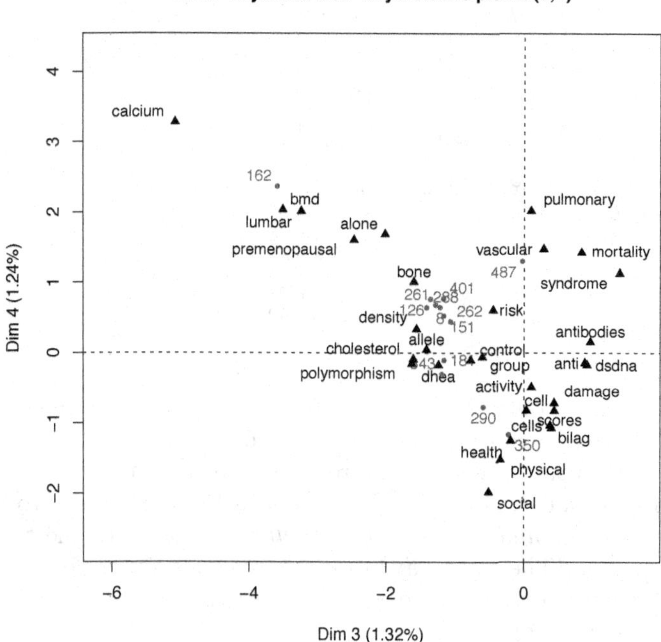

FIGURE 7.3
Meta-keys and doc-keys on the (3,4) factorial plane.

7.2.4 Analysis of the year-aggregate table

In the first part of the analysis, we studied the whole set of documents without taking publication year into account. Our second goal is to study how publications evolved over the 19 years. An ALT can be constructed using publication year as the aggregating variable. The resulting table has 19 rows and 786 columns. The number of publications per year is relatively stable:

```
1994 1995 1996 1997 1998 1999 2000 2001 2002 2003
  18   15   22   13   19   33   21   19   19   19
2004 2005 2006 2007 2008 2009 2010 2011 2012
  30   38   35   42   34   31   39   34   25
```

7.2.4.1 Eigenvalues and CA of the lexical table

CA of the ALT (Table 7.5) gives us the eigenvalues, which are also shown in Figure 7.4 in decreasing order. The percentage of inertia retained by the first factorial plane (Figure 7.5) is 19%.

TABLE 7.5
The first eigenvalues of the CA of the years × words table.

```
Correspondence analysis summary
Eigenvalues
       Variance % of var. Cumulative % of var.
dim 1      0.06     10.69                10.69
dim 2      0.05      8.25                18.94
dim 3      0.04      6.86                25.80
dim 4      0.04      6.71                32.51
dim 5      0.04      6.49                39.00
...
```

The first axis contrasts words related to standard treatments for lupus (left) to new treatments and gene polymorphism (right).

The second axis contrasts the word *dhea*, which is the anti-aging hormone used (controversially) as a therapy for certain forms of lupus, to words like *cells, cyclophosphamide, pregnancy,* and *illness*, related to the treatment of pregnant women. The word *sdi* for *system damage index*, found in a fairly central position, corresponds to a tool for evaluating organ damage for the most common form of lupus.

The years are spread out from left to right on the first axis, in such a way that chronology is roughly respected. In fact, three periods can be seen along the axis: a first one from 1994 to 2000, a second from 2001 to 2006, and a third from 2007 to 2012. The year 2007 corresponds to the first articles on biotherapies.

Recall that 1997 is the year with the least publications. However, it is also the year which most contributes to the inertia of the second axis, along with the word *cytotoxic* (see Figure 7.5).

7.2.5 Chronological study of drug names

In this section, we keep only the words corresponding to names of 48 drugs or lupus treatments. These words were selected by Mónica Bécue-Bertaut from information published on a lupus-related website. The resulting table has 19 rows, one for each year, and 48 columns. The resulting table has 19 rows, one for each year, and 48 columns.

The drugs used to treat lupus include (among others) nonsteroidal anti-inflammatory drugs (Ibuprofen, Advil, Motrin), corticosteroids (prednisone, methylprednisone), anti-malarials (hydroxychloroquine, chloroquine), immunosuppressants (cyclophosphamide—also called CS, azathioprine, my-

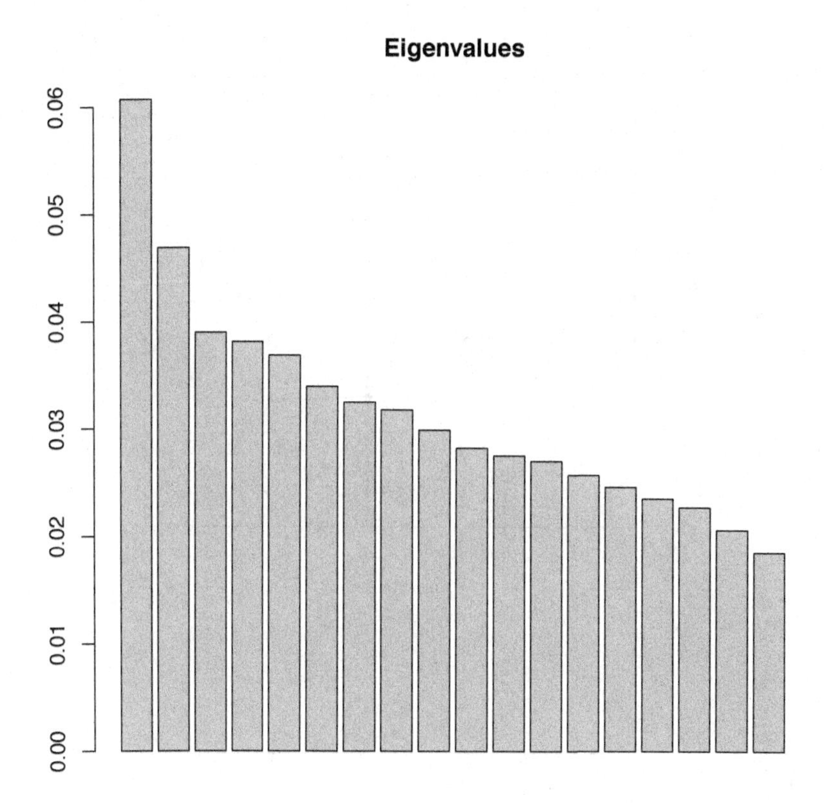

FIGURE 7.4
Eigenvalues of the years × words table.

cophenolate, methotrexate—also called mtx), monoclonal antibodies (rituximab, belimumab), and DHEA—also known as dehydroepiandrosterone.

Table 7.6 gives, by year, the names of the positive characteristic words (here, drugs) (test value > 1.96). Lupus is a complex disease that can affect different organs, which explains the variety of drugs and their evolution over the years. Note however that rituximab, classified as a biotherapy, appeared in 2002 and characterizes 2004 and the period 2006–2008. This drug is a cytotoxic monoclonal antibody that treats lupus but also other autoimmune diseases. From 2009 on, it was replaced by another drug of the same type, called belimumab.

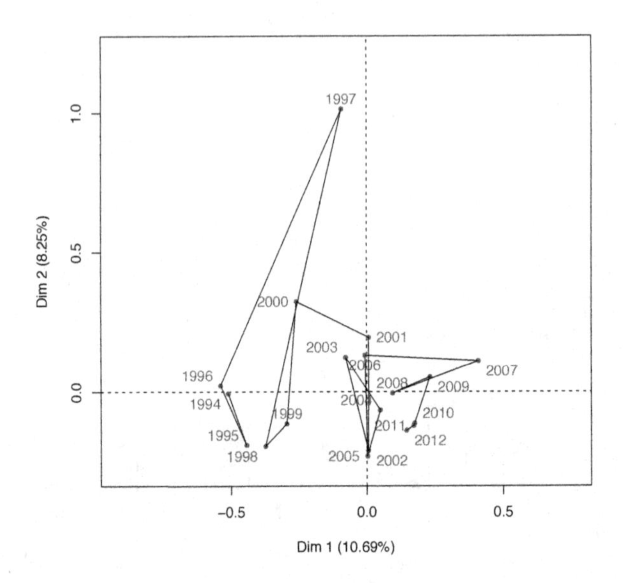

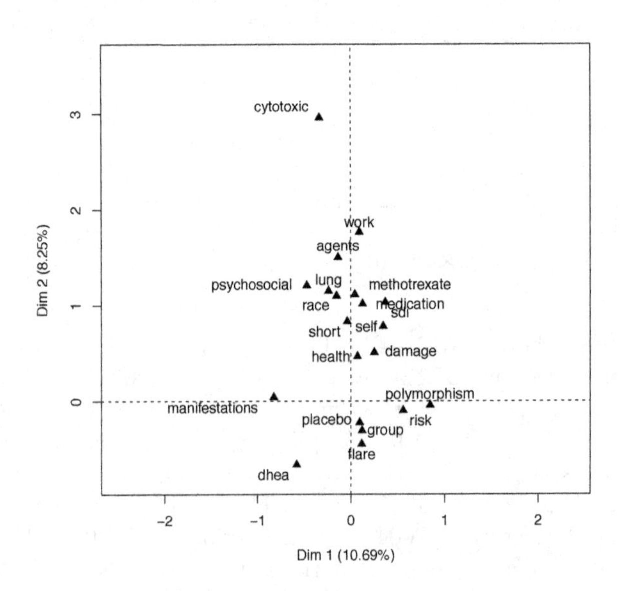

FIGURE 7.5
CA of the years × words table. Top: projection of the 19 years on the (1,2) factorial plane; bottom: projection of the words on the (1,2) factorial plane.

TABLE 7.6
Positive characteristic drugs for each year.

Year	Drug/treatment name (test value)
1994	corticosteroids(3.54), interleukin(2.01)
1995	prednisone(4.05), anticardiolipin(4.63)
1996	cyclophosphamide(3.27), anticoagulant(2.91), hydroxychloroquine(2.57), steroids(2.05)
1997	cytotoxic(7.46), methotrexate(3.36), steroid(2.55), corticosteroids(2.39), azathioprine(2.09)
1998	dhea(5.95)
1999	anticoagulant(2.24)
2000	prednisone(3.47), cyclophosphamide(2.95)
2001	antibodies(2.68)
2002	chloroquine(4.02), prasterone(2,61)
2003	cyclophosphamide(2.44), antibodies(1.98)
2004	dhea(3.55), prasterone(2.80), rituximab(2.51)
2005	antibody(3.45), prasterone(1.96)
2006	steroids(2.78), rituximab(2.73), chloroquine(2.11)
2007	rituximab(5.34), mm(2.48), immunosuppressive(2.22)
2008	prasterone(3.60), methotrexate(3.58), rituximab(2.35), belimubab(2.17)
2009	belimubab(4.01)
2010	dhea(2.35), plasma(2.31)
2011	belimubab(7.29), antibody(2.99), prednisone(2.15)
2012	belimubab(5.90), plasma(3.11), corticosteroid(2.74), mm(2.01)

7.2.6 Implementation with Xplortext

The code in `script.lupus.R` was used to obtain the previous results concerning the bibliographic database of scientific articles on lupus, and is available on the Xplortext.com website. The script is divided into three parts corresponding to the three earlier processing steps:

- **the first part of the script** involves the direct analysis of the abstracts:
  - the `TextData` function builds the (non-aggregate) LT; the corpus corresponds to column 5; only words found at least 15 times in at least 10 abstracts are retained; tool words are removed; only words with at least 3 letters are kept,
  - the `summary.TextData` function gives initial statistics on the corpus and the indexes of the first ten words,
  - the `LexCA` function performs CA on the LT; no plots are requested,
  - the `barplot` function plots the first ten eigenvalues,

  - the `summary.TextData` function lists the meta-keys and doc-keys,
  - the `plot.LexCA` function plots the meta-keys and doc-keys on the (1,2) and (3,4) planes.

- **the second part of the script** performs the chronological analysis:
  - the `TextData` function constructs the ALT (abstracts aggregated by year), and displays a small part of the resulting table,
  - the `summary.TextData` function provides some initial statistics on the corpus,
  - the `LexCA` function runs CA on the ALT; no plots are requested,
  - the `plot.LexCA` function plots, on the (1,2) plane, the years followed by the most-contributing words.

- **the third part of the script** involves the analysis of the years × drugs table:
  - the years × drugs ALT is constructed,
  - the `LexChar` function reports the characteristic drugs for each year, and the `plot.LexChar` function plots them.

7.2.7 Conclusions from the study

This example makes it possible to see how a bibliographical study progresses—one which does not aim to do every possible analysis. Despite the help provided by textual data software, data processing can be time-consuming. Indeed, the outputs generated by the factorial analysis are often extensive in the case of large databases. For this reason, practitioners may appreciate the help of meta-keys and doc-keys in identifying relevant topics. It is usually necessary to repeat the analyses for various frequency thresholds.

In a typical bibliographic search, researchers choose *a priori* several keywords that seem relevant to them. The use of such keywords for searching large databases like *Medline* makes it possible to identify huge numbers of references and abstracts that no researcher has time enough to read, many of which are not at all useful to them. An initial analysis like the one we describe here makes it possible to target topics related to a given practitioner's research. Meta-keys can be used to select a subset of interesting articles to be looked at in more detail.

A bibliographic study begins with the analysis of the table involving the original documents and a subset of the corpus's words. The next step is to move to an ALT. The aggregation criterion may be the year of publication, journal name, or any other relevant criterion available in the original body of work. It is possible to target a theme, leading to a reduced vocabulary size and an analysis involving a more detailed study of the chosen theme.

7.3 Badinter's speech: a discursive strategy

This section is the work of Mónica Bécue-Bertaut.

7.3.1 Introduction

Although CA is usually applied to large corpora made up of several documents, it can also be useful when studying a single long text with the aim of revealing its structure and the organization and progression of the argumentation. This is of particular interest in the case of a rhetorical speech, which aims to *convince* those that hear it. Here, the speech, which forms a continuous whole, is first cut into sufficiently long parts so that local variability, always present, can be smoothed out and structure can emerge in terms of a trajectory of the parts.

In this section, we give an example of this: CA applied to the speech delivered by Robert Badinter to defend the bill for the abolition of the death penalty in France. This lawyer was appointed Minister of Justice by François Mitterrand in June 1981. The repercussions of this speech, delivered on 17 September 1981 before the deputies of the National Assembly, helped to make known its author as one of the main actors in the defence of human rights. The duration of the speech was 62 minutes. The manuscript, handwritten, was later donated to the manuscripts department of the National Library of France and digitized. The speech can be downloaded from the official website `gallica.bnf.fr` in `.pdf` form.

7.3.2 Methods

7.3.2.1 Breaking up the corpus into documents

We thus download Badinter's speech in `.pdf` form and then simply save it in the `.csv` format. This automatically leads to a division into 124 document-paragraphs (= rows of the worksheet) whose lengths vary between 4 and 265 occurrences. First, in order to end up with sufficiently long documents, we create a contextual variable called *parts* which allows us to aggregate the paragraphs into 10 parts with 12 in each, and a final part with only 4. This arbitrary split is the starting point for our analysis.

7.3.2.2 The speech trajectory unveiled by CA

In a strongly structured argument-based text, we expect to observe well-differentiated but nevertheless connected parts because the arguments presented must be linked together, leading to a gradual renewal of the vocabulary. Therefore, by reordering the column words appropriately in the lexical table (here, parts $\times$ words), we would obtain a table with a *loaded diagonal.*

Note that the results of the CA are exactly the same, regardless of the order of the rows and columns in the analyzed table. Nevertheless, reordering either the rows or the columns (or both) would have the merit of revealing the structure of the data in the table itself. This point of view was developed by Jacques Bertin, who proposed a method to do this—unconnected to CA.

As we have seen, this structure is reflected in the CA results by an approximately parabolic trajectory of the parts. Of course, there are also irregularities and deviations from what would be the parabolic path, which constitutes, in a way, a reference model. Deviations, which may be atypical parts, backtracks, etc., may result from the oratory strategy of the author, or have some other meaning. This trajectory is considered to be the *geometrical form of the text* that captures its narrative and argument-based progression.

7.3.3 Results

The corpus size is 7977 occurrences and the vocabulary size is 1767 distinct words (see Table 7.7). The most frequent noun is the word *mort* (death, 104 times), followed by *peine* (penalty, 77 times) Next, after the auxiliary verb *a*/69 (has) and the adverb *plus*/50 (more), we find *abolition*/46. These 5 words are found in all of the 11 parts. The presence of the three nouns above in all parts shows that the theme of *l'abolition de la peine de mort* (abolishing the death penalty) appears regularly throughout the speech (see Table 7.8). This regularity endows this idea with great force. This is indeed a rhetorical technique.

An analysis of repeated segments supports this initial result. In effect, the most frequent segments among those retained are *la peine de* (the penalty of, 67 times) and *la peine de mort* (the death penalty, 65 times)—see Table 7.9. The most frequent repeated segments, arranged in alphabetical order, show how long segments give rise to multiple shorter segments. In this example, we have requested the indexes for both the words and the segments, arranged in both alphabetical and frequency order (not all of them are shown here). Indeed, it is useful to examine them from both points of view.

TABLE 7.7
Initial statistics: corpus size and vocabulary size before and after word selection.

```
TextData summary
                          Before   After
Documents                 124.00   11.00
Occurrences              7977.00 1674.00
Words                    1767.00  155.00
Mean-length                64.33  152.18
NonEmpty.Docs             124.00   11.00
NonEmpty.Mean-length       64.33  152.18
```

TABLE 7.8
Extract of the word index ordered by frequency.

Index of the 20 most frequent words

	Word	Frequency	N.Documents
1	mort	104	11
2	peine	77	11
3	a	69	11
4	plus	50	11
5	abolition	46	11
6	justice	40	7
7	parce	25	8
8	si	24	10
9	bien	22	10
10	aussi	21	10
11	pays	21	8
12	ceux	20	7
13	hommes	20	7
14	france	19	8
15	tous	19	8
16	là	18	6
17	homme	17	6
18	être	16	8
19	crime	15	7
20	autres	15	6

With the chosen thresholds, the parts × words ALT keeps only the 155 words used at least 5 times and found in at least 3 of the 124 document paragraphs. The CA of this table gives—which is one of the main results—the trajectories of the parts on the first factorial plane, retaining 31% of the total inertia. The first two eigenvalues are, respectively, 0.27 and 0.24, which suggests that a sufficiently large number of the 11 parts or aggregate documents contribute to the inertia of these first two axes. Cramér's V is equal to 0.40, which is high given that the largest it could be is 1. This indicates that the documents tend to use different vocabulary.

As is usual when the number of documents is small, we start by looking at their representation on the first CA plot (see Figure 7.6) without involving the vocabulary. The parts' labels provide information on chronology. The representation of the speech's trajectory over time provides a visual account of the rhythm of the lexical progression.

The overall pattern shows that the speech can be divided into four blocks [P1–P4], [P5–P7], [P8–P9], and [P10–P11], which are all fairly homogeneous since within each block, the parts are close together, even if not all of the confidence ellipses overlap. The trajectory of the first three blocks [P1–P4], [P5–P7], and [P8–P9], resembles quite clearly a parabola. The fourth block [P10–P11] marks a return towards the first. Between each of the first three blocks, there is a marked jump, indicating that a large part of the vocabulary in each is self-contained, whereas the first and last blocks share important words (see Table 7.10). As the confidence ellipses foresaw, this structure is

TABLE 7.9
Extract of the segment index in lexicographic order.

```
Number of repeated segments
  106

Index of the repeated segments
Number                              Segment  Frequency Long
69                              la peine de       67    3
70                         la peine de mort       65    4
71                   la peine de mort c est        4    6
72                     la peine de mort est        5    5
73                      la peine de mort et        4    5
74                       la peur de la mort        3    5
75                          la question de         5    3
76              la question de l abolition         3    5
77                              le droit de        3    3
78                             le plus haut        3    3
79                      les pays de liberté        3    4
80          mesdames messieurs les députés         4    4
81                              mort c est         6    3
82                        mort du coupable         3    3
83                              n a jamais         4    3
84                                 n a pas         4    3
85                               n est pas         8    3
86                                   n y a         5    3
87                            ne peut être         3    3
88                                où l on         3    3
89                           parce qu elle         6    3
90                             parce qu il         3    3
91                         pays de liberté         5    3
92                   pour la peine de mort         3    5
93  pour les partisans de la peine de mort         3    8
```

quite stable in relation to small variations in the word frequency threshold (not shown here).

7.3.4 Argument flow

Let us now see how the arguments used fit into this structure. We consider the words, supplemented by a few particularly interesting segments, as representing these arguments. The words plot can be superimposed onto the parts plot (see Figure 7.6). The word cloud is divided into four blocks—like the parts cloud is, which simplifies the study of the connections between both.

Only the most contributing words are retained for this plot, yet they summarize quite well the arguments used in the four blocks:

- In the first sequence [P1–P4], the Minister of Justice poses the question of the abolition of the death penalty with *question de l'abolition*, *de la peine de mort en*. Then, he recalls the efforts of several politicians, such as *Jaurès*, to move in this direction, as well as their *éloquence* (eloquence). Finally, he

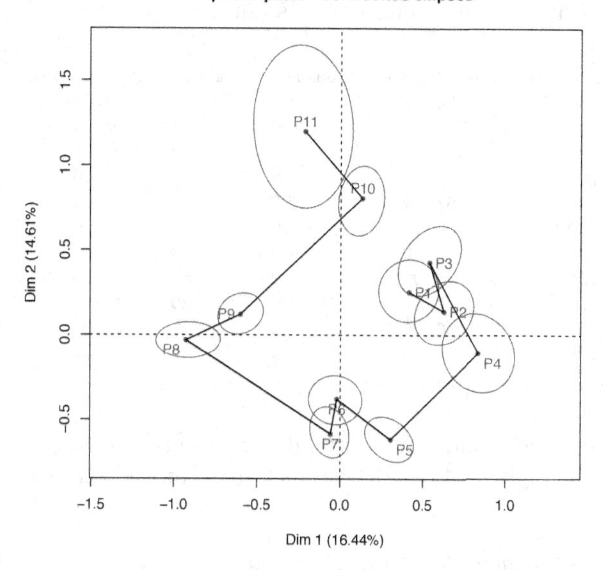

Speech parts and meta-keys

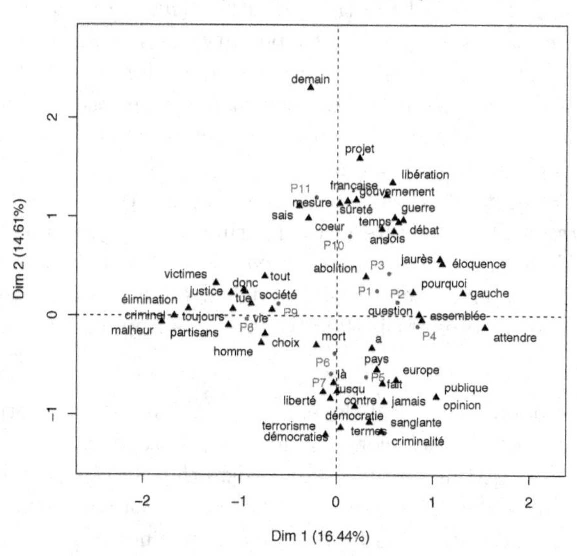

FIGURE 7.6
First factorial plane. Top: trajectory of the 11 parts of Badinter's speech. Bottom: parts and selected words (contribution over three times the average contribution).

TABLE 7.10
Extract of profiles of contributing words (sum per column = 100).

Docs	Terms éloquence	jaurès	opinion	criminalité	criminel	victimes	libération	débat
P1	16.67	20	0.00	0	0	0.00	0	25.0
P2	50.00	40	0.00	0	0	0.00	0	12.5
P3	33.33	40	0.00	0	0	0.00	40	0.0
P4	0.00	0	42.86	0	0	0.00	0	12.5
P5	0.00	0	57.14	80	0	0.00	0	0.0
P6	0.00	0	0.00	20	0	0.00	0	0.0
P7	0.00	0	0.00	0	0	0.00	0	0.0
P8	0.00	0	0.00	0	80	50.00	0	0.0
P9	0.00	0	0.00	0	20	33.33	0	0.0
P10	0.00	0	0.00	0	0	16.67	60	50.0
P11	0.00	0	0.00	0	0	0.00	0	0.0

asks himself what could justify *attendre* (waiting) any longer, citing the fear of *opinion publique* (public opinion) as a reason for France's delay in passing abolition.

- The second sequence [P5–P7] mentions at the beginning that one should not *refuser de se prononcer publiquement par peur de l'opinion publique* (refuse to speak out publicly for fear of public opinion), and again insists on France's lag in this area compared to other countries (*pays*, *europe*). The disconnect between *peine de mort* (the death penalty) and *démocratie* (democracy), as well as *liberté* (freedom) is emphasized as follows: *"dans tous les pays où la liberté est inscrite dans les institutions et respectée dans la pratique, la peine de mort a disparu"* (in every country where liberty is prescribed in its institutions and respected in practice, the death penalty has been abolished). Badinter argues that the death penalty has no dissuasive value against *criminalité sanglante* (violent crime): *ceux qui croient en la valeur dissuasive de la peine de mort méconnaissent la vérité humaine* (those who believe in the deterrent effect of the death penalty misunderstand human nature). He also notes its ineffectiveness in stopping *terrorisme* (terrorist acts).

- The third sequence [P8–P9] develops the theme of the horrific nature of the death penalty, which corresponds to *justice d'élimination* (justice by elimination). The Minister of Justice insists that *justice* (justice) cannot put a *homme* (man's) life into play. Abolition must not suffer exception, even in the case of *crimes odieux* (odious crimes). It is not a question of forgetting the *victimes* (victims), and he affirms that *"du malheur et de la souffrance des victimes, j'ai, beaucoup plus que ceux qui s'en réclament, souvent mesuré dans ma vie l'étendue"* (as for the misfortune and suffering of victims, I have throughout my life—much more that those that proclaim it—understood its immensity). He shows that *criminel* (criminal) and *victimes* (victims) share the *malheur* (misfortune). He points out that *justice qui tue* (justice that kills), *la justice d'élimination* (justice by elimination), is not *justice.*

- Lastly, the conclusion [P10–P11], a short final sequence, reuses certain terms from the first one, including *libération* (liberation), *française* (French), *débat* (debate), *temps* (time), *mesure* (measure), *guerre* (war), and *lois* (laws). This part reemphasizes the personal tone: *je sais* (I know), already very present throughout the speech (*je*—the personal pronoun *I*—is used 54 times), and insists in involving the parliamentarians: *vous* (you) and *votre* (your), a continuation of how these words were used throughout the whole speech (*vous* appears 50 times).

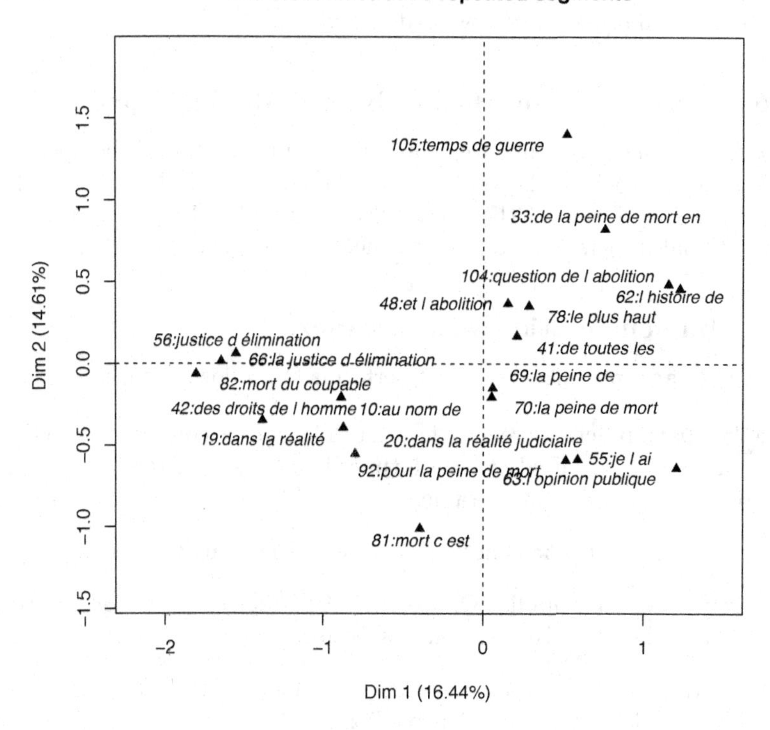

FIGURE 7.7
Representation of selected repeated segments on the principal plane of the CA.

The repeated segments (see Figure 7.7) show that certain very common and important words are used in multiple contexts. For instance, *justice* appears—to cite only the segments shown on the plot—in *justice d'élimination* (justice by elimination), *justice qui tue* (justice that kills), and *justice française*

(French justice). Depending on the distances between the different segments, it is possible to know if they are frequently used in the same parts (small distance) or not (large distance). This helps us to better understand their function in the arguments being made. In this example, *justice d'élimination* and *justice qui tue* are close together on the plot and can therefore be regarded as *synonymous expressions.* In contrast, the *justice française* segment deviates from the others as it is frequently used in different parts. Note nevertheless that we find *justice qui tue* and *justice française* together at the end of the speech, in the phrase: *"grâce à vous la justice française ne sera plus une justice qui tue"* (thanks to you, French justice will no longer be justice that kills), but this is precisely to insist on the fundamental difference that must exist between the two. Here, the repeated segments are useful in illustrating the factorial plots. This requires, however, a preselection since their large number excludes us from automatically representing all of them on the plot. Section 7.3.6 describes how to do this.

7.3.5 Conclusions on the study of Badinter's speech

The CA performed here uncovers progression in the text of the speech, and words associated with this progression. The methods we have used allow us to visualize the "form" of the text, aided by specific CA plots, and uncover organisational principles in terms of lexically homogeneous parts.

7.3.6 Implementation with Xplortext

The code in `script.ENBadinter.R` performs the following tasks:

- The `TextData` function creates the ALT. The corpus corresponds to the first column. Only words used at least 5 times in at least 3 of the 124 paragraphs (= source documents) are retained.
- The `summary.TextData` function provides initial statistics on the corpus.
- The `LexCA` function applies CA to the ALT. Segments are considered supplementary elements. No plots are asked for.
- The `ellipseLexCA` function represents the documents on the principal factorial plane, along with confidence ellipses.
- The `plot.LexCA` function represents the most contributive words on the principal factorial plane.
- The `plot.LexCA` function represents—initially—repeated segments whose representation quality on the principal factorial plane is at least 0.25 (selSeg="cos2 .25"). The choice of autoLab="yes" leads to a better visual display of the labels. From this initial representation (not shown here), we choose to retain only segments that do not overlap. For instance, for the

three segments: *de l'abolition* (abolishing), *question de l'abolition* (question of abolishing), and *la question de l'abolition* (the question of abolishing), we choose to keep only the second one. The choice of segments to retain is made in terms of their number (corresponding to alphabetical order). In this second call to `plot.LexCA`, we provide the vector of numbers of the selected segments: selSeg=c(10,19, ...).

7.4 Political speeches

This section is the work of Ramón Álvarez-Esteban and Mónica Bécue-Bertaut.

7.4.1 Introduction

The corpus of political speeches we examine here involves a small number (= 11) of long documents (around 10,000 occurrences in each). Textual data science methods—and the **Xplortext** package that implements them—can deal just as well with a corpus of a dozen or a hundred relatively long texts as they can with a corpus of a large number of short texts (like the majority of those seen earlier in the book). Here we focus more on presenting the steps to take and the choices made, rather than an in-depth interpretation of the results, which are closely related to the domain of political science in a country that may not be familiar to all readers. This corpus, with its slightly more than 100,000 occurrences, is the longest examined in the book.

The texts of speeches have been obtained from the published versions of the *Diario de Sesiones del Congreso de los Diputados* (*Journal of Sessions of the Congress of Deputies*), published by the Spanish parliament.

7.4.2 Data and objectives

We are interested here in the investiture speeches of Spanish presidential candidates made since democracy was reestablished. The date at which this occurred is generally considered to be that of the referendum for ratifying the Spanish Constitution, i.e., 6 December 1978. As a reminder, the long dictatorship of General Franco ended with his death on 20 November, 1975 at the age of 83.

The Spanish Constitution states that the King or Queen of Spain, after having consulted with representatives of the political parties that obtained seats in parliament, proposes a candidate as president. This individual must then give an investiture speech before the Congress of Deputies of Spain, describing the program of their future government, hoping to obtain a vote of confidence from the deputies. If the Congress accords them this confidence by an absolute majority (i.e., a favorable vote by at least half plus one of the deputies that make up the assembly—present or otherwise) the candidate becomes president. Otherwise, they must submit—in the following days—to a new vote of confidence, where a simple majority (more votes in favor than against by deputies present during a session in which a quorum is met) is sufficient. In the case of failure, another candidate must be proposed. If in the two months following the first vote, no candidate has obtained investiture by Congress, it is dissolved and new elections are called.

At the time this example was created, 11 speeches followed by investitures had occurred since December 1978. Typically, Spanish governments are stable, and only 2 of the 11 speeches correspond to the same Congress—the first one. In effect, Adolfo Suárez resigned on 29 January, 1981 under the combined pressure of all of the political actors of the time. The candidate for his succession, Leopoldo Calvo-Sotelo, did not obtain an absolute majority during his first vote of confidence on 20 February. He was undergoing the second one on 23 February when a military coup—quickly defeated during the night of 23 February—interrupted the investiture. The resumption of the session on 25 February, 1981 led to the election of Calvo-Sotelo with an absolute majority.

What are we looking for in the application of textual data science to these speeches? We are dealing with texts we understand well, occurred in a temporal reality we also know well. We can even situate the speeches in relation to a sequence of events whose dates are known exactly. What can we learn from a statistical analysis that we cannot get from simply reading or listening to a text? According to linguists, who have studied this is detail, *reception* of a text by reading or listening to it is different to its *emission*. We need to remain conscious that reading a text means already interpreting it implicitly—and not in an articulable way. In contrast, data science methods help to analyze *emitted* texts systematically, relegating interpretation to the results analysis stage. This is the point of view we adopt here.

The project leading to the construction of this corpus had the following objectives:

- Study whether what has been called the *Spanish transition to democracy*, or often more simply *the transition*, can be detected in the speeches. In general, this transition period is considered to be from the death of Franco to the first change in leading party in 1982 with the arrival in power of the Spanish Socialist Workers' Party (PSOE), led at the time by Felipe González.

- Find homogeneous and clearly demarked periods in terms of vocabulary use; find the boundaries of these periods and thus date changes between periods, and, perhaps, discover links between vocabulary and contextual variables (dates and candidate ideology).

It is a question of following the vocabulary flow in the succession of speeches, which are—in terms of development and context—very different to each other. These speeches are given (and prepared) by different people, with different ideologies, at well-separated times (a full term of Congress is four years), and deal with specific problems of the time. We can, however, suppose that they originate from the same *textual source*—in the words of André Salem—which is here a generic *Spanish presidential candidate*. Actually convincing the members of Congress is not an important goal since pacts—when necessary—have already been made, and the result of the vote is essentially known in advance. Nevertheless, the public image of these politicians is at stake in front of the public, made up of voters but in particular the media and specialists, who will comment on and analyze their speeches, increasingly with the help of data science methods.

7.4.3 Methodology

Here, CA is a fundamental method for displaying lexical similarities between speeches and identifying temporal relationships. Then, AHC with contiguity constraints (AHCCC) can help decrypt the factorial axes and detect changes and trends. Lastly, a description of the hierarchical tree's nodes can help understand the vocabulary flow.

What kinds of information do we want to obtain? First, we would like to display the trajectory of the speeches.

No matter the chronological corpus in question, vocabulary evolves with time. New words appear or their frequency increases, others disappear or become much less common. This *evolution* in vocabulary can happen at different rates depending on the corpus or its parts. It can happen at a regular pace, where—approximately—in each part of the same length the number of new words is constant, or speed up or slow down. The structure of this evolution is transmitted to the corpus, and as we have seen in Section 3.3.3, if evolution happens at a regular rate, the trajectory will have the shape of a parabolic crescent—in the words of Jean-Paul Benzécri. A comparison of this typical shape with the actual observed trajectory will provide information on the rate of evolution of vocabulary use, possible backtracking, sudden changes, and even lack of change. The AHCCC may confirm or provide further details on changes observed on the CA factorial planes. In addition, the dendrogram constructed using AHCCC uncovers the hierarchical structure of the set of speeches when the chronological dimension is taken into account. Labeling the tree using *hierarchical words* (see Section 5.3.3) can provide information on vocabulary flow. Above all, this labeling is able to distinguish *local* lexical characteristics—specific to each of the speeches—from those shared by longer sequences, i.e., made up of several time-contiguous speeches found in the same node. In this way, we can know which vocabulary characterizes long sequences overall, and distinguish it from characteristic words in each of the individual speeches.

7.4.4 Results

7.4.4.1 Data preprocessing

In order to retain the capital letters we included in the corpus when input from the *Journal of Sessions of the Congress of Deputies* and the semantic information they carry, the capital letters denoting the start of sentences have been manually removed from the database so as not to interfere with this. These retained capital letters serve in general to distinguish between homographs. Thus, *Gobierno* (the government) can be distinguished from *gobierno* (I govern). It is necessary to request to keep the capital letters directly in the code script.

7.4.4.2 Lexicometric characteristics of the 11 speeches and lexical table coding

To begin, we attempt to get an initial overview of the corpus without preselecting words. The corpus size is 101,967 occurrences and the vocabulary has 9416 distinct words (see Table 7.11). The respective lengths of the 11 speeches are shown as a bar chart in Figure 7.8. The longest speech is the first [Su79], given by Adolfo Suárez in 1979, with 12,149 occurrences. The shortest, with 7592 occurrences, is by Felipe González in 1989 [Gz89]. The 14 words used more than 1000 times each are all tool words (not shown here). The frequencies of the most-used full words are also shown in Figure 7.8. This plot is an output of the `plot.TextData` function, which allows stopwords to be removed before plotting. We see that the most-used full words are *política* (politics, 411 times), *Gobierno* (government, 403 times), and *España* (Spain, 320 times).

Next, we proceed to word selection as follows. We eliminate stopwords according to the list provided in the **tm** package, supplemented by the list (*consiguiente*, *ello*, *hacia*, *punto*, *Señorías*, *si*, *Sus*, *vista*, A, B, C, D, E, F, a, b, c, d). These words correspond to highly repeated expressions in individual speeches, corresponding to verbal tics like, e.g., *por consiguiente* (therefore), and terms used for enumeration (A, B, etc.). As for numbers, they are also removed. Finally, we keep only words used at least 10 times in at least 2 speeches. In this way, we end up with 32,181 occurrences of 1072 distinct words (see Table 7.11).

7.4.4.3 Eigenvalues and Cramér's V

We now apply CA to the LT (speeches $\times$ words). The main results are shown in Table 7.12. The strength of the connection between the speeches and the vocabulary, measured by Cramér's V, is 0.27 and thus average. As a comparison, this is below the value for Badinter's speech ($V = 0.40$, see Section 7.3), which itself is fairly high. However, the first two eigenvalues are small (resp. 0.13 and 0.10), though relatively larger than the following ones which—moreover—decrease in value slowly (see Table 7.12 and Figure 7.9).

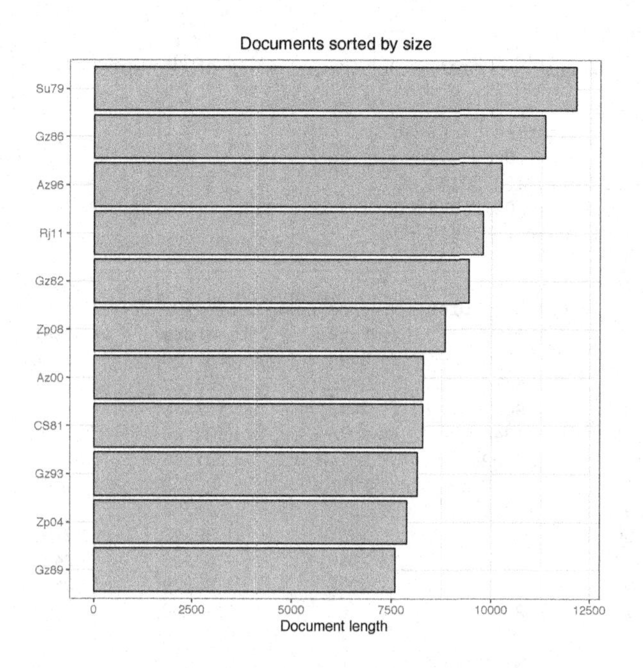

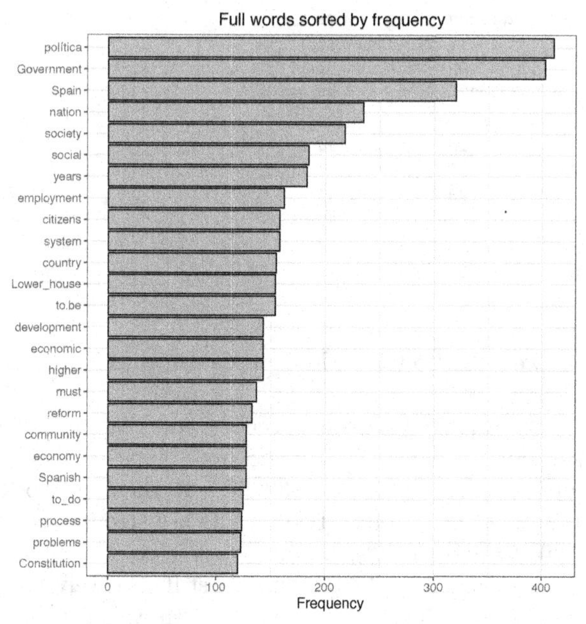

FIGURE 7.8
Initial description of the corpus. Top: bar chart of the initial length of each speech; bottom: bar chart of the frequency of the most-used full words.

TABLE 7.11
Summary of the lexicometric characteristics of the speeches.

```
TextData summary
                Before     After
Documents        11.00     11.00
Occurrences 101967.00 32181.00
Words          9416.00   1072.00
Mean-length    9269.73   2925.55

Statistics for the documents
   DocName Occurrences DistinctWords Mean Length100 Occurrences DistinctWords
                 before        before         before       after         after
1     Su79        12149          2825         131.06        3844           873
2     CS81         8274          2172          89.26        2594           778
3     Gz82         9427          2529         101.70        2707           799
4     Gz86        11344          2076         122.38        3609           780
5     Gz89         7592          1814          81.90        2430           735
6     Gz93         8141          2048          87.82        2640           778
7     Az96        10251          2352         110.59        3508           868
8     Az00         8287          1993          89.40        2797           772
9     Zp04         7882          2019          85.03        2443           705
10    Zp08         8833          2295          95.29        2647           755
11    Rj11         9787          2404         105.58        2962           790

Index of the  10  most frequent words
         Word Frequency N.Documents
1  política       411          11
2  Gobierno       403          11
3  España         320          11
4  Estado         235          11
5  sociedad       218          11
6  social         185          11
7  años           183          11
8  empleo         162          11
9  ciudadanos     158          11
10 sistema        158          11
```

Nevertheless, we will also take a look at the third axis and see what it brings to the interpretation.

A weak relationship between speeches and words, and small first two eigenvalues means, of course, that the first plane of this CA retains only a small proportion of the inertia (32.3%). It therefore corresponds to low representation quality. The first two factors are strongly correlated with the year (resp. −0.557 and −0.778), while the correlation of this variable with the third axis is much smaller (0.164), though still relatively larger than its correlation with the following axes (except the 7th with 0.15). This information will facilitate interpretation of the first axes as we know that they are related to the evolution of the lexical stock. We also know that temporal evolution is not the only phenomenon in play; we can expect to see other things appear in the following axes.

TABLE 7.12
Extract of the results summary for the CA of the speeches × words table.

```
Correspondence analysis summary

Eigenvalues
      Variance % of var. Cumulative % of var.
dim 1    0.128    18.276               18.276
dim 2    0.098    13.989               32.265
dim 3    0.076    10.798               43.063
dim 4    0.069     9.761               52.824
dim 5    0.064     9.142               61.967

Cramer's V  0.265    Inertia  0.702

DOCUMENTS
 Coordinates of the documents
      Dim 1  Dim 2  Dim 3  Dim 4  Dim 5
Su79 -0.067  0.567  0.017 -0.005 -0.224
CS81  0.101  0.332  0.006 -0.096  0.263
(...)

 Contribution of the documents (by column total=100)
      Dim 1  Dim 2  Dim 3  Dim 4  Dim 5
Su79  0.418 39.145  0.045  0.004  9.332
CS81  0.643  9.031  0.004  1.084  8.704
Gz82  0.335 15.072 14.782  6.431  0.059
Gz86 47.115  4.377  0.143 22.132 12.810
Gz89 21.220  3.656  0.898 36.102 11.070
Gz93  0.100  0.734  4.030  0.833 10.238
Az96  7.720  0.228 23.097  4.444  2.399
Az00  6.999  5.689  6.150  0.870  0.402
Zp04  4.278  0.971  5.368  2.567 14.470
Zp08  6.973 15.269  9.853 10.754 17.884
Rj11  4.199  5.828 35.631 14.779 12.632

 Squared cosinus of the documents (by row total=1)
     Dim 1 Dim 2 Dim 3 Dim 4 Dim 5
Su79 0.008 0.565 0.000 0.000 0.088
CS81 0.015 0.159 0.000 0.013 0.100
Gz82 0.007 0.242 0.183 0.072 0.001
Gz86 0.676 0.048 0.001 0.170 0.092
Gz89 0.367 0.048 0.009 0.333 0.096
Gz93 0.003 0.014 0.061 0.011 0.130
Az96 0.180 0.004 0.319 0.055 0.028
Az00 0.158 0.099 0.082 0.011 0.005
Zp04 0.094 0.016 0.070 0.030 0.159
Zp08 0.133 0.223 0.111 0.110 0.171
Rj11 0.082 0.087 0.410 0.154 0.123

SUPPLEMENTARY ELEMENTS
Coordinates of the supplementary quantitative variables
           Dim.1  Dim.2 Dim.3  Dim.4 Dim.5
year      -0.557 -0.778 0.164 -0.055 0.011
```

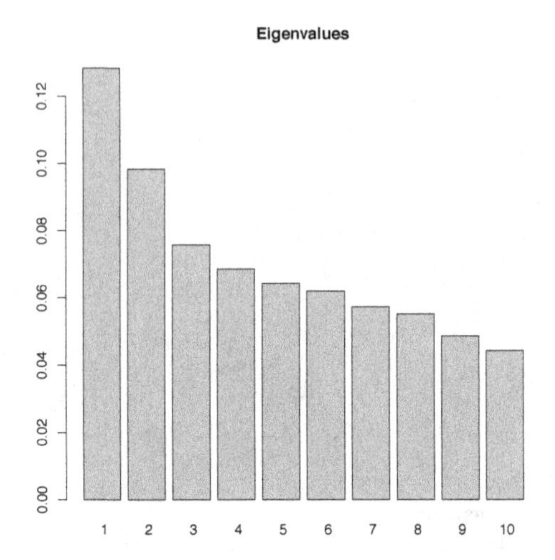

FIGURE 7.9
Bar chart of the eigenvalues.

7.4.4.4 Speech trajectory

The representation of the speeches (see Figure 7.10, top) on the (1,2) plane from the CA shows a tripolar structure. The three poles [Su79, CS81, Gz82], [Gz86, Gz89], and [Gz93, Az96, Az00, Zp04, Zp08, Rj11] gather temporally contiguous speeches, but also—as we will see below—ideologically heterogeneous tendencies in the case of the first and third poles.

The first pole corresponds to speeches given by actors in the democratic transition: *Suárez, Calvo-Sotelo*, and *González*, in the sense that their participation in politics began before Franco's death. An initial result is the following: in their investiture speeches, a tendency to have the same ideology does not translate to the use of the same words.

The second pole [Gz86, Gz89] corresponds to the two speeches with the largest positive coordinate values on the first factor, as well as the largest inertia values (very far from the center).

The last six speeches [Az96]–[Rj11], which make up more than half of the corpus, correspond to the third pole and are located at the end of the parabolic trajectory. These speeches are not very spread out on this principal plane and any coherent evolution seems to have stopped. To get a better idea of this, we can also take a look at the (1,3) plane.

We can refine this representation by drawing the temporal trajectory of the speeches. On the (1,2) plane from [Su79] to [Az96], this trajectory has—roughly speaking—the shape of a parabola whose axis of symmetry is parallel to the second bisectrix. The third axis gives an account of the evolution in

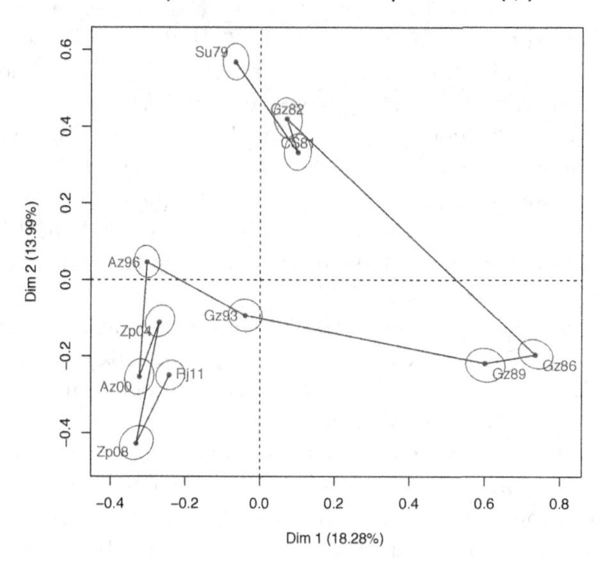

Ellipses around the documents points. Plane (1,3)

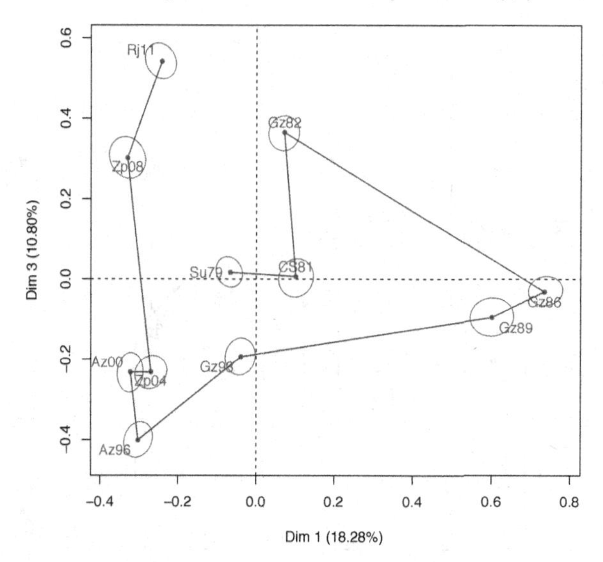

FIGURE 7.10
Representation of the speech documents and associated confidence ellipses. Top: (1,2) plane; bottom: (1,3) plane.

vocabulary over the most recent speeches, contrasting [Gz93, Az96, Az00, Zp04] with [Zp08, Rj11].

Here, this corresponds to a contrast between the latter two—which employ vocabulary linked to the financial crisis in 2007–2008, and the earlier ones, a contrast not related to ideology. The progression of these speeches can be seen along this third axis, which is—by construction—orthogonal to the principal plane. The fourth axis does not show any clear structure (not shown here).

The confidence ellipses drawn around each of the speeches are all small and do not overlap anywhere. If we increase the frequency threshold for words, the speech trajectory remains surprisingly similar, though the sizes of the confidence ellipses increases. We can therefore say that the structure of this set of speeches is stable.

It is illustrative to turn these two plots into a 3-dimensional representation (see Figure 7.11). Here, we clearly see the (approximate) shape of a parabola which visualizes the narrative connections existing between speeches. The trajectories we see on the plots point to the large difference between the vocabulary of [Gz93] and that of previous speeches from the same politician, perhaps due to the fact of having to govern without an absolute majority, and also due to the serious economic crisis that was just beginning at the time. They also highlight the meagre renewal of vocabulary brought by [Az96], [Az00], and [Zp04], which could be due to the poor oratorical abilities of these politicians.

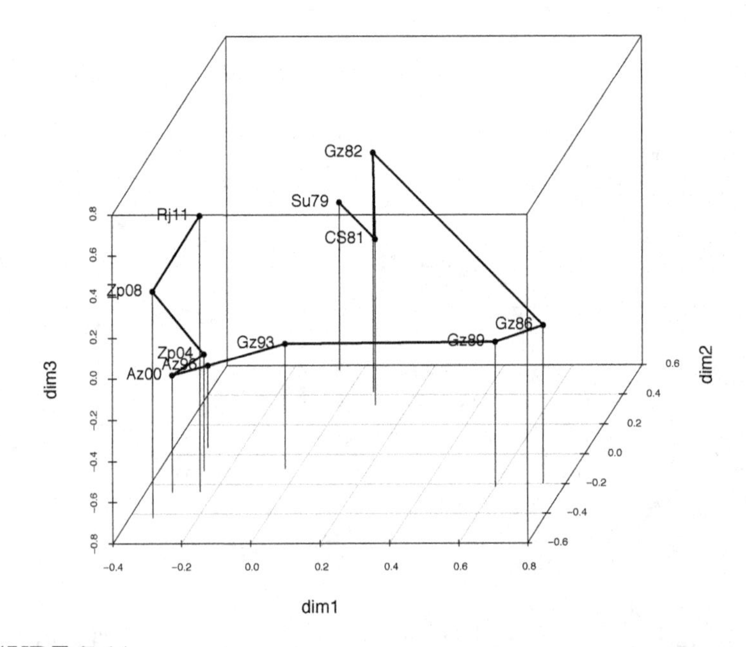

FIGURE 7.11
Three-dimensional representation of the points documents in the (1,2,3) space.

7.4.4.5 Word representation

The representation of the words on the factorial planes makes it possible to understand the nature of the temporal evolution in the speeches. On these plots we again find the tripolar structure that was seen in the speech cloud. The transition formulas between the document and word representations make it possible to define a temporal order to the corresponding three word poles. We will follow this ordering when decrypting the word representation on the (1,2) plane (see Figure 7.12).

The first pole, formed by words with coordinate values above 0.5 on the second axis, corresponds to the language of the transition. These words represent recent gains in democracy: *libertad* (liberty), *Constitución* (the Constitution), *Derecho* (the Law), etc.

Words in the second pole help understand the nature of the large change between the lexical profiles for [Su79], [CS81], and [Gz82] on the one hand, and [Gz86] and [Gz89] on the other. This change appears in the form of a large jump in the trajectory caused by the arrival of new words. Here, these words are mainly linked to the opening up of Spain to the rest of the world and its entry into the European Economic Community (accession to the EEC in 1985, official membership in January 1986). Note for example the words *Comunidad* and *Económica* (Economic Community) on the one hand, and *Acta* and *Única* (Single Act). Going back to the data uncovers that the first two have an association rate of over 1 with the speeches [Su79], [Gz86], and [Gz89], which means that these words are used more than average in those speeches (in terms of lexical profiles) (see Section 2.2.4 for the definition of *association rate*). The last two words, which allude to the Single European Act signed in 1986 by all of the members of what would become the European Union, are used more than average in only the speeches [Gz86] and [Gz89], which explains their excentric positions. Of course, besides *Comunidad* and *Económica*, the speeches [Gz86] and [Gz89] still share certain words with earlier speeches. For instance, *política* (politics), which is—let us remember—the most common full word in the whole corpus, is used more than average in the speeches from [Su79] to [Gz89], and correspondingly less than average in later speeches. In the same vein, *inflación* (inflation) is only associated with the speeches from [CS81] to [Gz89].

Reading off the words in the third pole is done by simultaneously looking at the (1,2) and (1,3) planes in Figure 7.12. In the (1,2) plane, we see rather what is shared by the speeches from [Gz93] to [Rj11]. On the third axis, we can look for the words which distinguish the earliest speeches from the last two: [Zp08] and [Rj11]. Hence, the (1,2) plane shows that all of the speeches from the third pole—from [Gz93] to [Rj11]—deal with *reforma* (reform). Sometimes, the type of reform necessary is mentioned, e.g., reforming the Constitution starting from [Zp04] to solve the question of the territorial integrality of the Spanish state, and—from [Zp08] on—as a solution to massive unemployment.

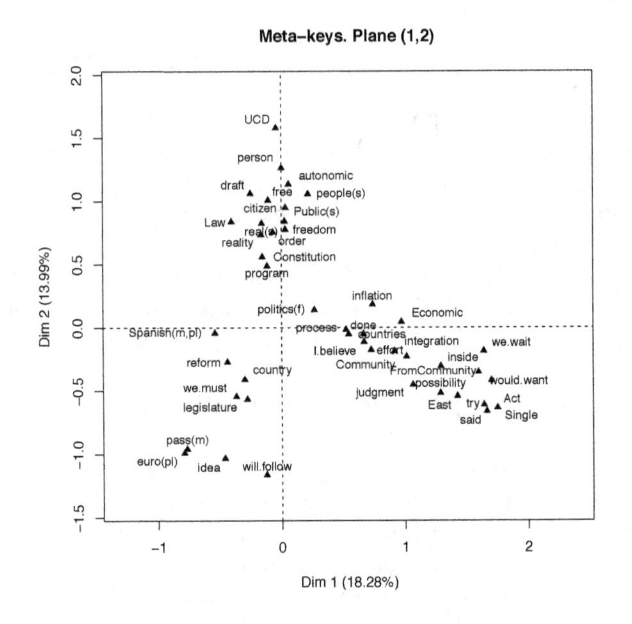

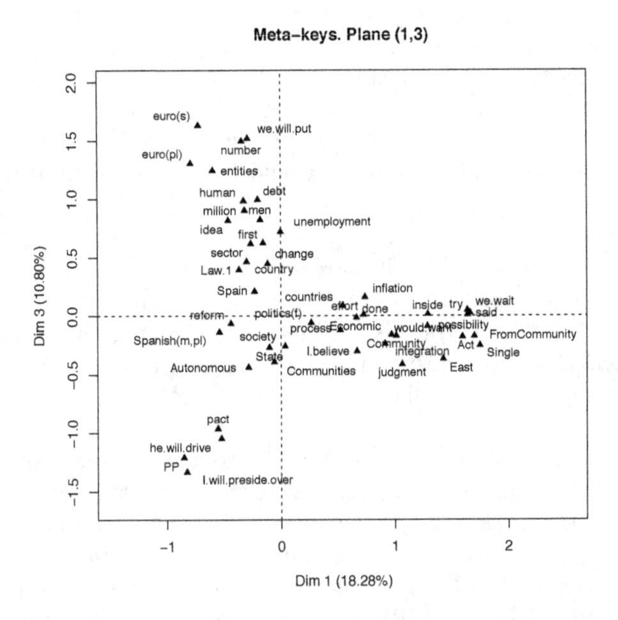

FIGURE 7.12
Representation of words with contributions over 6 times the average on at least one or both axes. Top: words contributing to axes 1 and 2; bottom: words contributing to axes 1 and 3.

However, this word *reforma*, often used in a rather vague way, also seems to act as a substitute for talking about a clearly defined government agenda.

As for the third axis (Figure 7.12, bottom), it shows which words act to differentiate the speeches [Gz93]–[Zp04] from [Zp08] and [Rj11]. The former seem to suffer from a certain verbal poverty, with only the words *pacto* (pact), *Comunidad*, and *Autónoma* (Autonomy) appearing, highlighting the fact that three of the four governments in question had to make pacts with regional parties due to an absence of an absolute majority. We note that in these speeches, the word *Comunidad* is associated with *Autónoma*. In the last speeches, [Zp08] and [Rj11], the economic crisis forces them to mention it, though with reluctance—especially [Zp08]. We thus see words like *euro*, *euros*, *deuda* (debt), *millones* (millions), and *paro* (unemployment) appear.

7.4.4.6 Remarks

After an initial period from [Su79] to [Gz82] marked strongly by the political transition, the second period from [Gz86] to [Gz89] corresponds to the Socialists' arrival in power and Spain joining the EEC, with two speeches far removed from the rest. From 1993 on, i.e., normalized relations with the rest of the world and the framework for autonomous regions established, we see little change in vocabulary in the final six speeches, especially from [Az96] to [Zp04]. This is quite intriguing. Decrypting the meaning of such small changes in vocabulary would best be done in the presence of a political scientist. However, we can already make several comments:

- It is probably necessary to relativize the weak temporal vocabulary renewal of the last six speeches compared to the much more significant influx of new words in the first five—from [Su79] to [Gz89]—which correspond to the immediate post-Franco period. During that time, many changes occurred, leading to a rapid evolution.

- Furthermore, in the years 1979–1989, political leaders had to guide and push changes in certain directions, which forced them to put clear temporal signals in their speeches that situated them in relation with the past and the future.

- The evolution in vocabulary from [Gz93] onwards occurs along the third axis, orthogonal to the principal plane. This signifies that this evolution occurs using a set of words that has little to do with words used in earlier years. For example, the *euro* did not exist before 2000, and little was said about *deuda* (debt) before [Zp04]. As for *transparencia* (being transparent), though mentioned in [Su79], it only started to be used frequently from [Gz93] on.

- When an important event occurs like the serious economic crisis in 2008, it leads to the appearance of new vocabulary; this corresponds well to the big jump we see in the trajectory along the third axis starting in 2008, and even more clearly in 2011. The 2007–2008 *crisis* pushed candidates to react in terms of the words used in their investiture speeches, even though the word *crisis* is never actually used in this period.

- The last six speeches differ little from each other, at least as far as their general overviews of proposed government policies are concerned, even when their ideologies differ. Each speech focuses on specific themes, which might appear on later axes.

7.4.4.7 Hierarchical structure of the corpus

An AHCCC, operating on the three axes containing interpretable information from the speeches, can give us another point of view on the same data. The hierarchical tree obtained (see Figure 7.13) is coherent with the three poles detected on the principal plane (= division into three clusters) and with the division of the most recent speeches' pole into two subpoles on the third axis (= division into four clusters).

We also note that AHC itself also leads to these same three or four clusters, thus serving as a validation of these partitions as robust against the type of method chosen. However, as we might have expected, AHC does not always respect the temporal ordering at each step. Indeed, the node [Gz93, Az96, Az00, Zp04, Zp08, Rj11] is aggregated with the non-contiguous node [Su79, CS81, Gz82] before [Gz86, Gz89] is joined with them when creating the final node (the corresponding hierarchical tree is not shown here). The temporal contiguity is not always respected at the lowest level of the tree.

Labeling the tree with manually added *hierarchical words* (see Section 5.3.3 for the definition), as shown in Figure 7.13, provides another view of vocabulary changes. This labeling has the great advantage of differentiating what is specific to a speech from that which corresponds to larger time periods. In order to obtain a visually interpretable tree, we only show the most highly characteristic hierarchical words.

Table 7.13 reports the hierarchical words associated with the node [Su79,CS81,Gz82], as output by the `LabelTree` function.

Let us briefly summarize the tree from top to bottom. Two large time periods are identified: [Su79–Gz89] and [Gz93–Rj11]. The first is dominated by *política* (politics), the second by *reforma* (reform), whether it be of the Constitution, the labor code, the administration, structural, or unspecified. This time period can itself be divided into two clusters, corresponding to two poles already identified by CA: [Su79–Gz82] and [Gz86–Gz89]. Then, lower down, the second large time period divides in the same way as the third axis of the CA was seen to split into two contrasting parts. These tripolar and—at a lower level, quadripolar—structures are clearly the dominant structural elements here. They guide our interpretation of the results, leading to a straightforward summary of the lexical characteristics of the corresponding time periods.

Thus, it is clear that:

- The period [Su79–Gz82] uses the language of the transition, with words like *pueblo* (people), *hombre* (man), and *dignidad* (dignity).
- The period [Gz86–Gz89] is dominated by words related to integration into the EEC.
- The final period [Gz93–Rj11] is that of *reforma* (reform), *competitividad* (competitivity), and *innovación* (innovation).
- The division of the final period into two subperiods shows how the first one [Gz93–Zp04] (a rather economically successful period—at least from 1996 onwards) limits itself to language related to the pacts required with regional parties. The second, [Zp08–Rj11], deals with economic issues though not in any precise way.
- If we drop down to the level of the individual descriptions of [Zp08] and [Rj11], the former does not really tackle the economic crisis. The words *euros*, *millones* (millions), and *economía* (the economy), are associated with the node [Zp08–Rj11]. However, the words individually associated with [Zp08] are not related to the economy, while those characterizing [Rj11] are strongly linked to the economic crisis.

The labeled tree in Figure 7.13 helps us to identify lexical characteristics of this sequence of speeches, and contrasts it globally with each of the individual speeches.

TABLE 7.13
Extract of the list of hierarchical words associated with the node containing [Su79], [CS81], and [Gz82].

```
round(res.Label$'Su79 CS81 Gz82',2)
                       Intern % glob % Intern freq Glob freq  p.value v.test
pueblo/ people(s)          0.23   0.08          34        45        0   6.49
palabras/ words            0.10   0.03          15        18        0   4.66
autonómico/ autonomous 0.11   0.04          16        20        0   4.62
crisis/ crisis             0.22   0.11          33        59        0   4.39
Pública/ Public            0.12   0.05          18        25        0   4.38
programa/ program          0.30   0.17          45        91        0   4.25
ciudadana/ citizen         0.11   0.04          17        24        0   4.17
dignidad/dignity           0.09   0.04          14        19        0   3.92
real/ real                 0.11   0.05          17        26        0   3.79
hombre/ man                0.06   0.02           9        10        0   3.79
pueblos/ people(pl)        0.09   0.04          14        20        0   3.70
Proyecto/ Draft            0.09   0.03          13        18        0   3.68
histórica/ historical  0.08   0.03          12        16        0   3.67
```

7.4.4.8 Conclusions

In this example, we have obtained some quite interesting results:

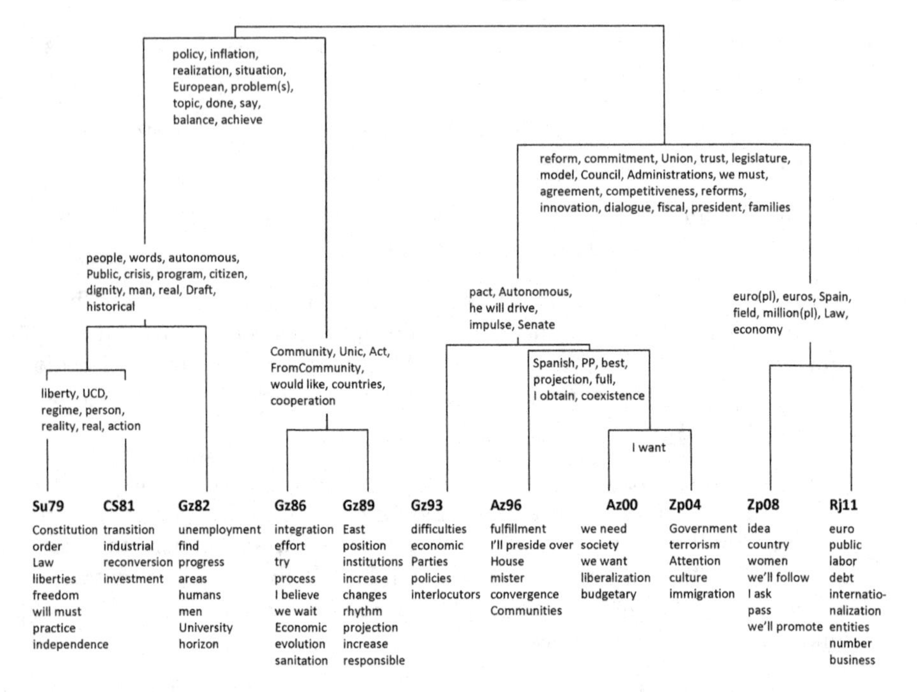

FIGURE 7.13
Hierarchical tree with contiguity constraints.

- The first change occurs with [Gz86] and not [Gz82] as would have been expected. The arrival in power of the PSOE did not lead to marked changes in the investiture speech, which instead happened once the left consolidated power.

- Three very distinct periods are visible, as we saw above. These correspond to different epochs, even though each of them contains ideologically heterogeneous speeches.

- In the speeches we examined, ideology is not a factor that differentiates vocabulary usage.

- The trajectory of the final speeches shows little evolution over time. Each of them deals with its own distinct themes (e.g., economic liberalization for Aznar in 2000, terrorism for Zapatero in 2004, violence against women for Zapatero in 2008).

The entirety of these results should be provided to the end-user(s) (political scientists, sociologists, linguists, etc.) and discussed and interpreted with them. In this way, it is possible to obtain different points of view on what the results mean, which could lead—for instance—to the necessity of further studies. In the present corpus, each of the speeches could be studied separately as was done for Badinter's speech (see Section 7.3). This corpus could certainly also provide more information on the possible decline in the quality of public speaking in Spain, related to—,in the words of the Spanish historian Julián Casanova: *"the interruption of parliamentary life between 1936 and 1977, which disconnected Spanish politicians from this type of speech, something which did not occur in other Western democracies"*.

7.4.5 Implementation with Xplortext

Results were obtained using the lines of code in `script.discours.R`.

- The function `TextData` is runned two times in order to obtain the initial statistics and plots on documents and words that are wanted. For the second run, the tm and a user stoplists are taken into account.
- Then, the function `TextData` is runned once again considering the argument values to be used in the CA. The corresponding LT is built. Only words used at least 10 times and in at least 2 speeches (=source documents) are kept. The tm and a user stoplists are taken into account.
- The LexCA function is applied to the LT. The summary.LexCA function offers the main numerical results and the meta-keys.
- Several plots are displayed representing the eigenvalues, documents (enriched by confidence ellipses and the chronological trajectory), words (the most contributing ones).
- A tridimensional representation of the documents, enriched by their chronological trajectory, is also offered.
- Finally, a contiguity constrained hierarchical tree is built on the speeches, from the coordinate values on the first three axes. CA is runned once again, with the same argument values as before, except for `ncp=3` in order to keep only three axes in the clustering step. The nodes are described by the hierarchical words.

7.5 Corpus of sensory descriptions

This section is the work of Sébastien Lê and Mónica Bécue-Bertaut.

7.5.1 Introduction

Presentation of the corpus of sensory free comments analyzed in this section requires first a few words on what *sensory analysis* and *sensometrics* are.

To obtain an objective measure of the sensory properties of a set of products, *quantitative descriptive analysis* (QDA) and its variants have been widely used by the food industry. The general principle is the following. A tasting sheet, i.e., a list of descriptors, is established, covering the essential sensory aspects of the tested products (*acidity, bitterness*, etc.). Then, trained *judges* score each descriptor for each product to establish its *sensory profile*. The scores thus collected then give rise to different statistical analyses, the simplest of which is to compare the means. The set of statistical methods for analyzing tasting or "sensory" data constitutes the discipline known as sensometrics.

Collecting and analyzing tasting sheets is a proven methodology and remains the basis of sensory descriptions which claim measurement status. However, it misses something: nothing is said about the relative importance of the different descriptors in the perception of the products (for example, the fact that one taster places a lot of importance on bitterness, another on acidity). This observation is not a new one, and to overcome this problem, some propose to directly and globally conceptualize the differences between products. Methods designed for this purpose (called *holistic methods*) include *categorization*, also known as *sorting* task. Categorization requires judges to partition the set of products, i.e., cluster together products found to be similar (according to their own criteria). To interpret such results, it is very useful to collect, during the *hall test session*, verbal descriptions of the clusters thus formed. We then speak of *verbalization task*. It is possible to directly analyze the corpus thus obtained to determine and characterize differences between the products. This is what we do here in this section, by analyzing the verbal part collected to enrich a categorization task.

7.5.2 Data

7.5.2.1 Eight Catalan wines

In this section, we study data collected during a hall test session of Catalan wines. Our main goal is to show why verbal categorization can be of interest. In this context, the choice of wines is not fundamental in itself, and merely helps obtain sensory responses. Here we perform a secondary analysis of these data in order to present an original methodology for analysing only the verbal part.

Based on wine that was available at the time, eight red wines from Catalonia were selected according to a full factorial design involving the following three factors with two levels each:

- Two *designations of origin* or DO wines (Priorat and Ampurdan, labeled P and A). The Priorat DO is in fact "DOC", i.e., *qualified designation of origin*. Only Priorat and Rioja wines are DOC in Spain.
- Two *dominant grape varieties*: Grenache or Carignan (also called Samsó in Catalan and Cinsault in English), denoted G and S.

- The two *vintages* 2005 and 2006 (denoted 05 and 06).

For a given varietal and DO, the wines of both vintages come from the same producer. These wines were chosen by Joan-Miquel Canals from the Faculty of Enology at Rovira i Virgili University (URV, Tarragona).

In the following, the wines are identified by an acronym that summarizes the levels of the three factors. For example, the Priorat DO wine made principally from Grenache grapes from the 2005 crop is identified by PG05.

7.5.2.2 Jury

In this type of study, a jury or panel of judges—,either experts or "naive consumers", tests a set of products. The number of judges is low when experts are involved (perhaps 12), but high in the case of naive consumers (e.g., 100). The main objective is to describe the products. We also seek to know how acceptable they are and the degree of pleasure or displeasure they provide. In the example analyzed here, a jury of 15 judges working in France (wine experts: enologists, sommeliers, and salespeople) blindly tasted the eight wines described above.

7.5.2.3 Verbal categorization

The tasting method chosen was the sorting task (i.e., categorization). The protocol is as follows. Judges are asked to categorize the I products tested into at least 2 and at most $I-1$ ($=7$ here) clusters based on their own sensory criteria. This is the categorization step itself. The division into clusters reflects perceived similarities/differences between products without actually providing the criteria used to do so—which generally differ from one judge to the next. In order to identify these criteria, we ask the judges to describe each cluster with freely chosen words. This leads to a fairly particular—and usually quite short—corpus.

In addition, the judges attribute a liking score from 1–10 to each wine. In the following, we call the mean across the 15 judges for each wine the "French score". A panel of 9 Catalan experts also evaluated the same wines using the same protocol. Here, we will only use their mean scores—denoted the "Catalan score".

7.5.2.4 Encoding the data

We shall use the notation from the MFACT section in Chapter 6 as this is the method used to analyse the results. For each judge $t, \{t = 1, \ldots, T\}$, an *individual table* $\mathbf{Y}_t$ is constructed, with one row per wine (I rows) and as many columns as there are *individual words*, i.e., the set of words used by that judge (J_t columns). The words associated with a given group of wines are considered to describe all of the wines in that group. These are presence/absence tables: the entry y_{ijt} at the intersection of row i and column (j, t) is equal to 1 if the group to which wine i belongs to was described using the individual word (j,t) by judge t, and 0 otherwise.

Next, the multiple contingency table $\mathbf{Y}$ of size $I \times J$, where $J = \sum_t J_t$, is constructed by binding row-wise the tables $\mathbf{Y}_t$. Usually, the total number of individual words J is greater than the number of different words used by the set of judges since several judges may use the same word, and each such usage corresponds to a separate column in $\mathbf{Y}$. Furthermore, individual words used by different judges may correspond to the same word in a given language, called a *global word* in this context. If so, we say that these individual words are *homologous.* Two other tables playing supplementary roles are also constructed and binded to the previous one: the quantitative table with two columns corresponding to the French and Catalan scores, and the sum table, i.e., the crosstab wines $\times$ global words. This stocks the frequency with which each global word is associated with each wine, irrespective of which judge used the word.

7.5.3 Objectives

We have the following objectives:

- Extract structure shared across as many individual configurations as possible in order to provide an average configuration for the wines based on their sensory proximities and contrasts.
- Identify the criteria governing proximities.
- Determine *consensual words*, i.e., words used by judges to describe the same wines, which may suggest that they are describing the same sensory perceptions.

7.5.4 Statistical methodology

7.5.4.1 MFACT and constructing the mean configuration

Formally, we could analyze each table using a *separate CA*, which would provide the *individual configuration* of the wines corresponding to each judge. As for MFACT, it makes it possible to directly obtain a mean configuration of the wines that is the closest possible to this set of individual configurations. Furthermore, MFACT balances out the influence of the different tables in the construction of the mean configuration, which is useful here because we want each judge to have *a priori* the same influence. In the representation obtained, two wines are close together if they are close together in most of the individual configurations.

7.5.4.2 Determining consensual words

In what we have seen so far, homology between individual words is not taken into account. When it is, the following question arises: do homologous words have the same meaning for all of the judges, i.e., do they correspond to the same sensory perceptions? To determine what are known as *consensual* words, we can use a method proposed by Belchin Kostov, Mónica Bécue-Bertaut, and François Husson corresponding to the `WordCountAna` function in the R package **SensoMineR**. We consider that the meaning of homologous words is shared when they are associated with the same wines.

Let us briefly present the method here. In the representations of wines and individual words given by MFACT, the more two wines are described using the same individual words, the closer they are, and analogously, the more two individual words describe the same wines, the closer they are. Consequently, the more homologous individual words are associated with the same wines, the closer they are together, and when close, are considered to correspond to the same sensory perception, i.e., are called consensual.

The starting point for determining whether a homologous word is consensual or not is the following. Suppose we have a global word corresponding to m homologous individual words. We first calculate the coordinates of the CoG of these m individual words, then the within-class inertia of this global word's subcloud, which is used as a statistic to measure the level of consensus for this global word. This value then needs to be tested. Here, the null hypothesis corresponds to a lack of consensus, i.e., an association between one word and different wines, according to the judges. To examine this hypothesis, we can establish the distribution of the within-class inertia when the m individual words are chosen at random from the set of such words. By doing this a great number of times (10,000 by default), the observed value of the within-class inertia can be situated with respect to the distribution of this null statistic and a p-value can be calculated. The value of this is equal to the proportion of samples of size m drawn at random that have a within-class inertia less than or equal to that of the word being examined. This test is performed for all words used by at least 3 judges (the default value in the function).

7.5.5 Results

7.5.5.1 Data preprocessing

When inputting the data, words corresponding to the same sensory perception are grouped together. For example, *acide* (acidic) can be used to cover all of *acide*, *acidulé* (tart), and *acidité* (acidity). Furthermore, certain composite words like *de garde* (for ageing) are turned into simple words by removing the space. In addition, tool words have been removed.

Then, finally, words obtained in this way are translated into English using dictionaries specializing in wine vocabulary. This does not modify the structure of the data on the condition—of course—that no two words are translated to the same English word.

7.5.5.2 Some initial results

The judges used between 4 and 21 individual words each. Some of these words are used by several of them (like *woody*), others by just one (like *pencillead*). Overall, there are 165 individual words but only 86 distinct ones, of which the most frequent are *woody* (37 times), *fruity* (36 times), and *tannic* (31 times). The corpus size is 410 occurrences. The active multiple table, binding the 15 individual tables row-wise, has 8 wine rows (or documents) and 165 column words. Here, a document corresponds to the set of occurrences associated with a given wine. The sum table, used as a supplementary table, has 8 row documents and 86 column words. The score table, also supplementary, has 8 row documents and two columns of quantitative variables.

7.5.5.3 Individual configurations

Each judge provides a multidimensional wine configuration that can be visualized by peforming CA on the corresponding individual table (see Figure 7.14). For all except three of them, the first factorial plane retains at least 70% of the inertia and therefore provides a good-quality visualization. The plot at the top of Figure 7.14 shows overlapping words and wine labels, due to the fact that groups are described, except for one word, by proper vocabulary.

7.5.5.4 MFACT: directions of inertia common to the majority of groups

To directly obtain the mean configuration, we apply MFACT to the multiple table binding the T individual tables, the sum table, and the score table. In this multiple table, there are as many active groups as there are judges. The sum table and score table are considered supplementary tables.

The inertia of the first factor is equal to 8.3 while the maximum value, equal to the number of active groups (i.e., judges) is 15. This means that the first global axis of the MFACT does not correspond to the first direction of inertia in each of the 15 individual configurations. Indeed, it must be remembered that sensory perceptions vary greatly from one individual to another—even between trained judges. It is for exactly this reason that we work with a panel of judges and not simply one expert. Furthermore, the categorization criteria are different for different judges, as they were not explicitly told what criteria to use. Nevertheless, the correlation coefficient between the first factor and the projection of the 15 configurations on this axis, called the *canonical correlation coefficient* in MFACT, is over 0.70 for 9 of the judges (see Table 7.14). Thus, this direction of inertia is indeed present, with at least average intensity in the majority of individual configurations.

In the absence of a validation technique, this value of 0.70 is chosen empir-

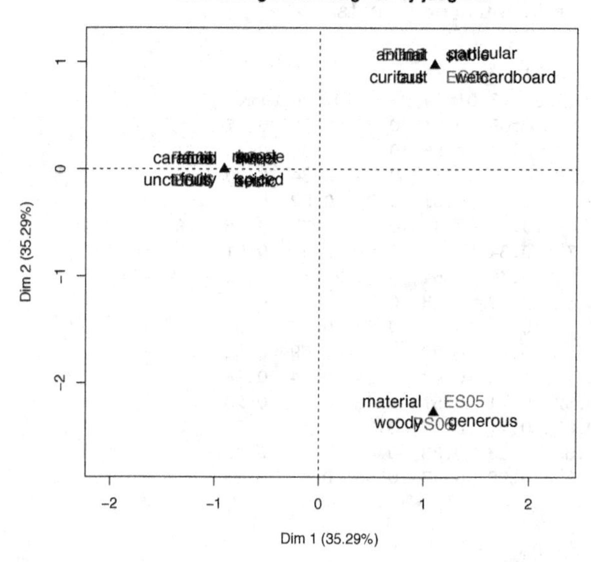

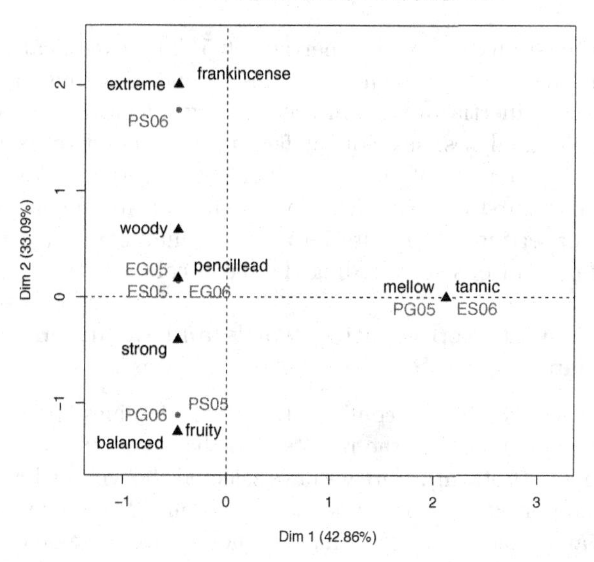

FIGURE 7.14
Examples of individual configurations. Top: the configuration corresponding to judge E5 who categorized the wines into 3 clusters. Bottom: the configuration corresponding to judge E12 who categorized the wines into 4 clusters.

TABLE 7.14
Canonical correlations coefficients.

```
> res.mfact$group$correlation
    Dim.1 Dim.2 Dim.3 Dim.4 Dim.5 Dim.6 Dim.7
E5   0.93  0.90  0.68  0.51  0.29  0.56  0.17
E6   0.62  0.77  0.73  0.56  0.75  0.63  0.44
E12  0.95  0.56  0.97  0.75  0.26  0.34  0.26
P1   0.90  0.84  0.72  0.61  0.26  0.52  0.32
P2   0.46  0.76  0.80  0.43  0.27  0.32  0.46
P3   0.96  0.73  0.34  0.73  0.51  0.53  0.51
P4   0.62  0.93  0.79  0.64  0.19  0.55  0.33
P5   0.96  0.63  0.62  0.53  0.76  0.51  0.22
P6   0.26  0.73  0.11  0.07  0.62  0.07  0.01
P7   0.96  0.84  0.58  0.92  0.57  0.69  0.51
P8   0.72  0.12  0.14  0.49  0.54  0.24  0.58
P9   0.67  0.52  0.71  0.57  0.63  0.12  0.23
P10  0.93  0.83  0.24  0.63  0.51  0.36  0.59
P11  0.90  0.59  0.13  0.33  0.62  0.35  0.43
P12  0.62  0.92  0.98  0.17  0.46  0.67  0.29
```

ically as follows. The canonical coefficients corresponding to the highest-rank axes are supposed closely to what would be seen when there is no structure. Here, for instance, all the values corresponding to the 6th or 7th axis are less than 0.70.

As for the second factor, its inertia is 6.5. The canonical correlation coefficients are over 0.70 for 10 judges. This second axis thus also corresponds to a direction of inertia present in the majority of individual configurations. Continuing the analyses, subsequent factors correspond to axes of inertia in only a minority or none of the individual configurations. We shall therefore focus only on the first plane, which contains 46% of the inertia—which is usual in this type of setting (see Figure 7.15). The remaining 54% corresponds to a diversity of perceptions surrounding the mean configuration.

7.5.5.5 MFACT: representing words and documents on the first plane

Figure 7.15 (top) shows the configuration of the 8 wines on the first MFACT plane. An interpretation of the results must be based on the wines that contribute the most (bottom). Here we have selected the wines whose contribution is greater than or equal to 1.5 times the mean one. The four wines retained—which here are those whose coordinate values are extreme on one or the other of the two first axes, are organized into three poles: PS06/ES05, PG06, and EG05. The first two poles are opposite each other on the first axis, and both in turn opposite the third pole on the second axis. These contrasts are neither associated with a region nor a grape variety. It is therefore necessary to use the words to explain the axes.

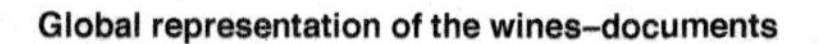

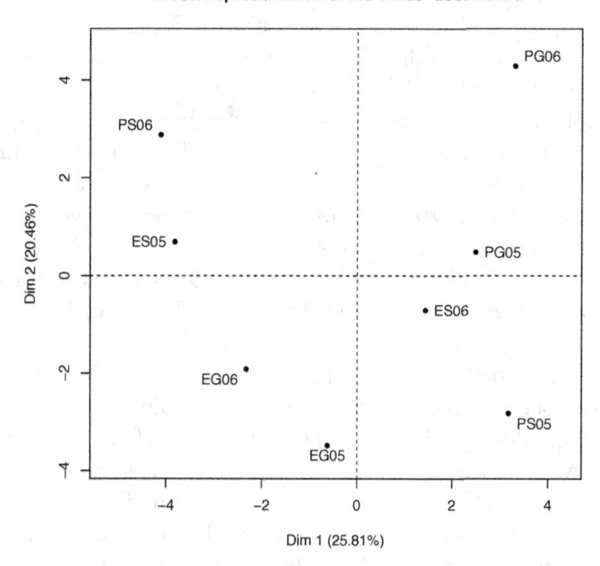

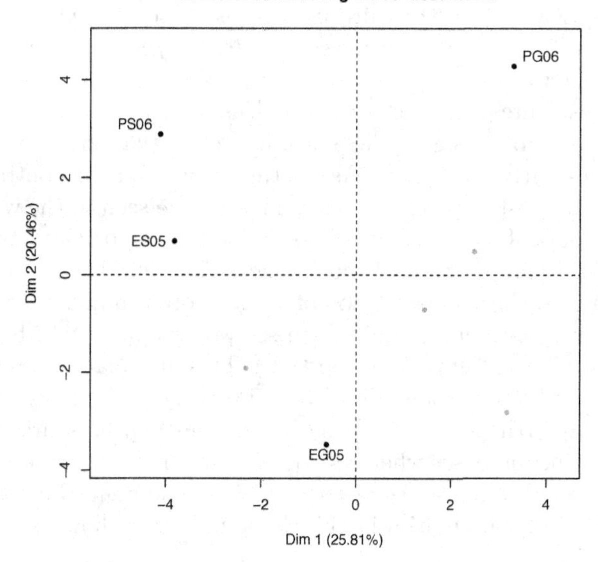

FIGURE 7.15
MFACT (1,2) plane: representation of the documents. Top: all of the documents; bottom: most contributive documents.

7.5.5.6 Word contributions

Because of the transition formulas, the representations of individual (active) and global (supplementary) words (see Figure 7.16) make it possible to describe differences and similarities between wines in terms of their sensory characteristics.

The homologous individual words (active columns) are mapped one by one on to the plots. In order to make the representation more readable, only words whose contribution is equal to or greater than twice the average contribution have been plotted (see Figure 7.16). A strong consensus seems to exist among the judges as to the meaning of most words since often, the homologous individual words (= same global word) are close to each other. For example, this is the case for *woody*, *woody.1*, *woody.2*, etc. Therefore, they are associated with the same wines.

We find in the words the tripolar structure found in the wine configuration, except for the set of words close to *powerful*. The wines from the first pole (PS06/ES05) are described by *woody* or other words frequently associated with this type of wine (*vanilla, candy*, etc.). These wines can also be considered *mellow*. If we also include words with weaker contributions, ones related to *for ageing* and other words related to wooded characteristics appear. The wine PG06 (from the second pole) is *fruity, ripe, supple, balanced,* and *round*. Note that this pole involves more words that the *woody* one; indeed, the *fruity* characteristic, less precise than *woody*, needs to be nuanced by other words. As for the wine EG05 (from the third pole), it is associated with rather negative features such as *astringent, aggressive, bitter, drying*, and *acid*, expressed in many different ways.

From these three poles, characterized using the most extreme wines, we can describe the axes as a whole. Thus, it is clear that the first axis contrasts wooded wines with fruity ones. As for the second axis, it contrasts defective wines with the most appreciated ones. This can be seen in the words found in the positive part of the axis, as well as in the strong correlations between the French and Catalan scores and the second axis (equal to 0.90 in both cases).

Furthermore, the representation of global words summarizes—in a certain way—the word representation. We can see words appear which confirm what was seen above—see Figure 7.16 (bottom). This first plane gives a clear image of the wines, whether wooded or fruity, and appreciated or not. These two main traits are orthogonal, which is quite interesting. It is often thought that the price of wine increases when aged longer in barrels. It therefore seems that such ageing is not a guarantee of quality, except no doubt for certain wines—which our limited capabilities in this domain do not allow us to characterize well.

The wine and word (individual active words or supplementary global ones) configurations in the (3,4) plane (not reproduced here) show that there is little consensus among the judges on the contrasts seen on these axes. The group

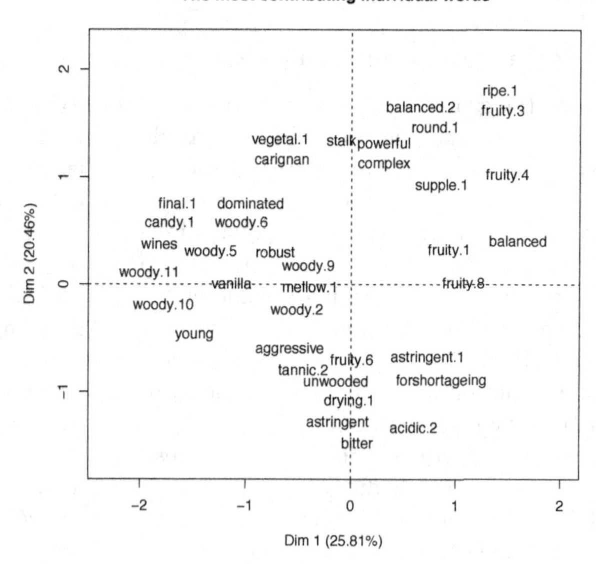

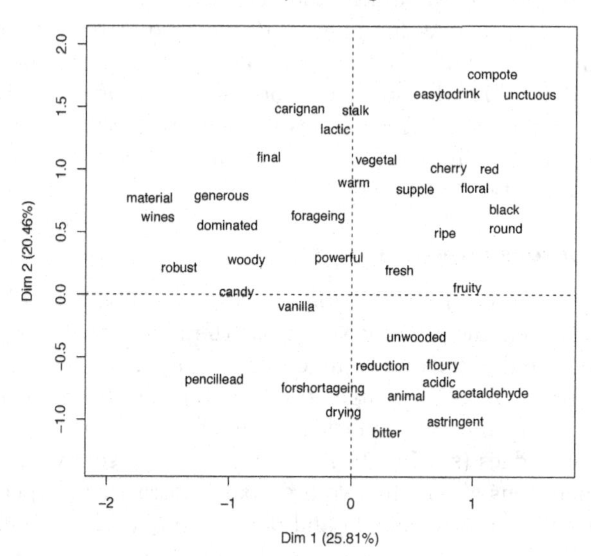

FIGURE 7.16
Word representation. Top: most-contributing individual words (active columns). Bottom: best-represented global words and the word "unwooded" (supplementary columns).

representation (not shown) suggests that both axes 3 and 4 are related to the specific behavior of certain judges.

7.5.5.7 MFACT: group representation

In this example, the group representation corresponds to that of the judges—for the active groups, and the sum and score tables—for the supplementary groups. First let us say that this representation (see Figure 7.17) makes it possible to see that (a) there is what can be described as an "average" level of consensus between judges, (b) the global directions of dispersion of the MFACT are directions of inertia strongly present in the sum table, especially in the case of the first axis (coordinate value of "SumTable" almost equal to 1, the maximum possible value), and (c) the second axis is strongly related to the liking scores. Point (b) justifies the use of column word sums as a summary of the individual column words in terms of their projections onto the MFACT axes (supplementary columns).

If we focus on a particular judge, this representation gives information about the presence or not of directions of inertia corresponding to the first two axes of their individual configuration. Thus, we can see whether each judge gives importance or not to the woody characteristics of a wine and/or finds it acceptable or not (and how strongly). For example, the rather high coordinate values of judge E5 on both axes shows that this judge gives great importance to both the woodiness and acceptability of wines (see Figure 7.17 and Figure 7.14—top). As for judge E12, the judge gives great importance to the woodiness of wines (see Figure 7.17). In the individual configuration provided by judge E12, the opposition between fruity and woody corresponds to the second axis, along which wines range from *fruity* to *extreme* (coming from *extreme woody*) and *frankincese* passing by the word *woody* along the way (see Figure 7.14—bottom).

7.5.5.8 Consensual words

By following the procedure described in Section 7.5.4.2, we obtain the consensual words, i.e., those associated with the same wines over most of the judges. Given the descriptive nature of this exercise, it is legitimate to use the value of 0.1 for the p-value. Thus, five consensual words are identified (*woody, candy, long, animal, fruity*), which is a satisfactory number considering the low number of judges (see Figure 7.18). These consensual words reinforce our previous conclusions about the two first axes contrasting, respectively, woody to unwooded perceptions (axis 1) and defects to qualities (axis 2).

7.5.6 Conclusion

Starting with the heterogeneous individual configurations, MFACT can be used to construct the mean configuration of the wines. In this particular case, the latter differs little from that obtained by CA applied to the sum table. This

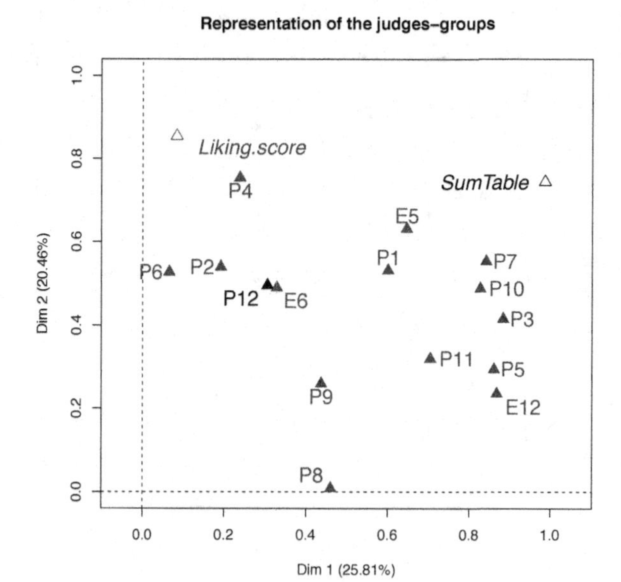

FIGURE 7.17
Representation of the groups-judges.

is because the words are associated with the clusters and—for each judge—reflect the structure over the wines induced by the proposed clusters.

Nevertheless, even here, it is very useful to apply MFACT due to the broader range of results it provides, in particular the representation of the judges and the detection of consensual words. In addition, with MFACT it is not necessary to choose a minimum frequency threshold, which would otherwise be necessary when an infrequent word is too influential in determining a certain axis.

In addition, we must emphasize that the methodology used here can be applied even in cases where:

- The judges used different languages, without performing translation, except for when determining consensual words. This is of great advantage in international case studies—which are increasingly common.
- The words are directly associated with the products, which is very useful when looking for appropriate words to describe new products. Various methods can be used to generate such words. Using MFACT to process this type of textual data can help uncover descriptors both shared by most of the judges and discriminating among the different products.

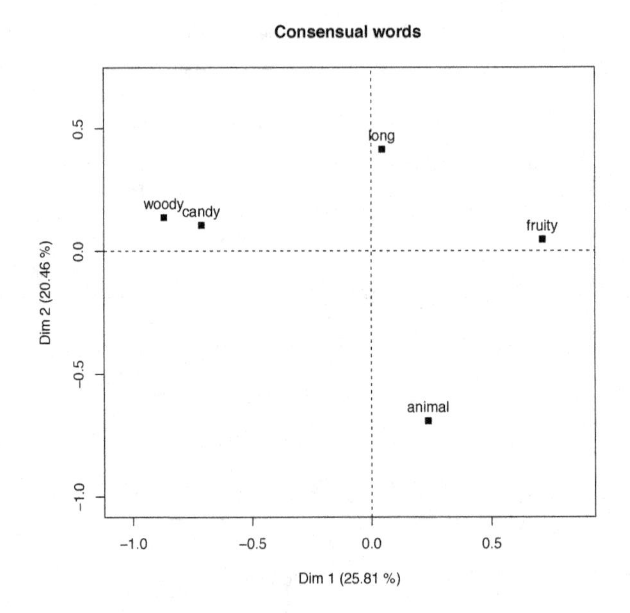

FIGURE 7.18
Representation of the consensual words.

7.5.7 Implementation with Xplortext

The code in `script.EnDegustation.R` performs the following tasks:

- First, iteratively, the fifteen individual tables are separately built and analyzed using the functions `TextData` and `LexCA`. The corresponding corpora correspond, respectively, to columns 1 to 15. All the words are retained. As they are built, these tables are binded row-wise to form a multiple table. As examples, the individual configurations provided by judges E5 and E12 are plotted. The table summing all the comments on the wines is then built and binded row-wise to the former multipe table. The same is done for the two columns filing the liking scores.
- The MFACT function is applied to the resulting multiple table. The main numerical results are listed.
- The `plot.MFA` function displays, successively, the global representation of all the documents, only the most contributing documents, only the best represented documents, the most contributing words, the best represented words and, finally, the judges (=groups).
- The `WordCountAna` function, included in the **SensoMineR** identifies the consensual words (argument proba=0.1) and plots them.

Appendix: Textual data science packages in R

The R software is a free tool for computational and graphical statistics. Built collaboratively, it is distributed freely by CRAN (Comprehensive R Archive Network). A huge number of add-on packages have been written for the R environment by the statistics community. Each such package is generally dedicated to a "family of statistical methods". R and associated packages can be downloaded at http://cran.r-project.org and installed on all usual operating systems: Windows, Mac OS, and Linux.

The Xplortext package

The **XplorText** package associated with this book is written for the R environment. It helps to perform textual data analysis using multidimensional statistics, following the path of Jean-Paul Benzécri and Ludovic Lebart. The heart of the package is correspondence analysis (CA; the `LexCA` function) and clustering methods for objects in factorial coordinates (AHC; the `LexHCca` function). The package can also hierarchically cluster documents under contiguity constraints (the `lexCHCca` function). The nodes of a hierarchical tree constructed using this function can have *hierarchical words* added to them (see Chapter 5), which can help the study of vocabulary flow. The `LexChar` function extracts *characteristic words* and *documents* from parts of a corpus. In addition, it is possible to use various functions from the **FactoMineR** package to perform a sequence of analyses, e.g., multiple factor analysis (`MFA`) and correspondence analysis for generalized aggregated lexical tables (`CAGALT`), as briefly decribed below. The `MFA` function implements the MFACT method and thus allows the simultaneous analysis of several contingency tables. The applications of this are numerous and varied in the textual domain; in particular, the analysis of multilingual corpora and the joint analysis of replies to different open-ended questions. As for the `CAGALT` function, not applied in this book, it jointly analyzes textual and contextual data, with the latter playing an interpretative role. **Xplortext** has its own website (http://Xplortext.org) offering tutorials and other tools.

As usual with R, the **Xplortext** package needs only to be installed once, then loaded whenever it is required with the command:

```
> library(XplorText)
```

Other textual data science packages

Other statistical analysis of texts packages are available in R. We briefly mention those which complement **Xplortext**.

The package **RcmdrPlugin.temis** by Milan Bouchet-Valat is based on the same classical methods as **Xplortext** but does not offer the most recent methods described in this book. However, it does allow for the analysis of co-occurrences, calculation of similarity measures between documents, and the evolution of the use of one or several words over time. It can analyze texts imported from spreadsheets, e.g., Alceste, Twitter requests, data from Dow Jones Factiva, LexisNexis and Europress, and folder systems of plain text files.

The **tm** package of Ingo Feinerer, Kurt Hornik, and David Meyer provides a great number of tools for constructing lexical tables. Some of its functions are even used by **Xplortext** and **RcmdrPlugin.temis**.

The package IRaMuTeQ created by Pierre Ratinaud is not an R package, but can interface with it for analyzing textual data. It allows textual data in specific formats to be read in, and can then generate scripts that can analyze the data in R. Developed in the LERASS laboratory, it provides classical methods such as contingency table construction, CA and clustering, but also outputs visually interpretable plots of co-occurrences, and moreover can perform divisive hierarchical clustering similar to that in the Alceste software package of Max Reinert. Information on IRaMuTeQ can be found at http://www.iramuteq.org, including documentation (in French) by Lucie Roubère and Pierre Ratinaud, and a step-by-step tutorial by Elodie Baril and Bénédicte Garnier.

Bibliography

The bibliography is limited to some essential books. It is divided into references about the exploratory multivariate statistical methods that are applied (correspondence analysis, clustering methods and multiple factor analysis) and those about textual data science. Two references appear in both parts. We have favored books written in English, although some background books written in French by Jean Paul Benzécri, Ludovic Lebart and Escofier & Pagès cannot be ignored.

Exploratory Multivariate Statistical Methods

- Benzécri J.P. (1973). *L'Analyse des Données, Tome I Taxinomie, Tome II L'Analyse des Correspondances.* Dunod, Paris.
- Benzécri J.P. (1992). *Correspondence Analysis Handbook.* Marcel Dekker, New York.
- Blasius J. & Greenacre M. (2014). *Vizualization and Verbalization of Data.* Chapman & Hall/CRC Press, Boca Raton, Florida.
- Escofier B. & Pagès J. (2016). *Analyses Factorielles Simples et Multiples: Objectifs, Méthodes et Interprétation*, 5th ed. Dunod, Paris.
- Gower J.C. & Hand D.J. (1996). *Biplots.* Chapman & Hall/CRC Press, London.
- Husson F., Lê S. & Pagès J. (2017). *Exploratory Multivariate Analysis by Example Using R*, 2nd ed. Chapman & Hall/CRC Press, Boca Raton, Florida.
- Lebart L., Morineau A. & Warwick K. (1984). *Multivariate Descriptive Statistical Analysis.* Wiley, New York.
- Lebart L., Piron M. & Morineau A. (2006). *Statistique Exploratoire Multidimensionnelle. Visualisation et Inférence en Fouilles de Données*, 4th ed. Dunod, Paris.
- Murtagh F. (2005). *Correspondence Analysis and Data Coding with R and Java.* Chapman & Hall/CRC Press, Boca Raton, Florida.

- Pagès J. (2015). *Multiple Factor Analysis by Example Using R.* Chapman & Hall/CRC Press, Boca Raton, Florida.

Textual Data Science

- Benzécri J.P. (1981). *Pratique de l'Analyse des Données, Tome III, Linguistique et Lexicologie.* Dunod, Paris.
- Blasius J. & Greenacre M. (2014). *Vizualization and Verbalization of Data.* Chapman & Hall/CRC Press, Boca Raton, Florida.
- Lebart L., Salem A. & Berry, L. (1998). *Exploring Textual Data.* Kluwer, Dordrecht.
- Murtagh F. (2005). *Correspondence Analysis and Data Coding with R and Java.* Chapman & Hall/CRC Press, Boca Raton, Florida.

Index

Printed in the United States
by Baker & Taylor Publisher Services